VHF, Summits and More

VHF, Summits and More

Having Fun With Ham Radio

BOB WITTE, KØNR

SIGNAL BLUE LLC
MONUMENT, CO

VHF, Summits and More
Copyright © 2019
Signal Blue LLC
All Rights Reserved.

ISBN: 9781795613194

Cover: Bob/KØNR on the summit of Mount Sneffels. (Photo: Joyce Witte, KØJJW)

Contents

Acknowledgements ix

1. Have Fun With Radios — 1
2. Frequency, Band and Wavelength — 3

Part I. VHF Operating

3. VHF FM Operating Guide — 7
4. Getting Started on 2m SSB — 21
5. Six Meters: The Magic Band — 25
6. The Myth of VHF Line of Sight — 29
7. How to Work A VHF Contest — 35
8. So You Want To Be a Rover — 41
9. VHF QRP — 49

Part II. Blog Posts

10. Choose Your 2m Frequency Wisely — 59
11. FOT: Frequency, Offset and Tone — 67
12. VHF Grid Locators — 73
13. Rescue on Uncompahgre Peak — 77
14. The Use of 146.52 MHz — 83
15. Paul Rinaldo's Rule — 87
16. Don't Get Stuck On 2 Meters — 89
17. VHF Distance From Pikes Peak — 91
18. Yes, Band Plans Do Matter — 95
19. Radio Hams are Not First Responders — 97

20.	Amateur Radio is Not for Talking	99
21.	The History of Electronic Communications	101
22.	Seventy Three	103
23.	Proper Kerchunking	105
24.	A Simple Wilderness Protocol	107
25.	Can I Use My Ham Radio on Public Safety Frequencies?	109
26.	Phonetic Alphabets	113
27.	Twisted Phonetic Alphabet	117
28.	Go Ahead and Use Phonetics on 2m FM	119
29.	Go Ahead and Call CQ on 2m FM	123
30.	Religion and Ham Radio	127
31.	Sorry, I've Been On 2m FM Again	129
32.	What's In Your Rubber Duck?	131
33.	A Better Antenna for Dual-Band Handhelds	137
34.	Is the Internet Destroying Amateur Radio?	139
35.	That's Not Real Ham Radio	145
36.	Amateur Radio: Narrowband Communications in a Broadband World	149
37.	We've Got Some Explaining to Do	155
38.	VHF FM: The Utility Mode	159
39.	Pursue Radio Operating Goals	163
40.	The VHF Digital Cacophony Continues	167

Part III. Summits On The Air (SOTA)

41.	How To Do A VHF SOTA Activation	175
42.	How To Do a SOTA Activation On Pikes Peak	179
43.	Mt Herman: SOTA plus VHF Contest	183
44.	The Ten Essentials for Hiking (and SOTA Activations)	187
45.	The Most Radio-Active Mountain in Colorado	189
46.	Yaesu FT-1DR: A Trail Friendly SOTA HT	191

47.	SOTA plus NPOTA on Signal Mountain	195
48.	Monarch Ridge South SOTA Activation	199
49.	Smoky Mountain SOTA	203
50.	Pikes Peak SOTA Winter Activation	211
51.	More Power For VHF SOTA	215
52.	Operating Tips for VHF SOTA	221
53.	VHF/UHF Omni Antenna for SOTA Use	225
	Glossary	231

Acknowledgements

Special thanks to my wife, hiking partner and soulmate, Joyce Witte (KØJJW), for her outstanding assistance with this book.

The following people provided valuable feedback on the book concept and contents: Bob Garwood (WØBV), David Remy (WAØKXO), Dan Romanchik (KB6NU) and Stu Turner (WØSTU). George Palecek (ABØYM), Paul Signorelli (WØRW), Joyce Witte (KØJJW) and Ken Wyatt (WA6TTY) provided photographs used in this book. Thank you so very much for your assistance!

1. Have Fun With Radios

When people ask me what ham radio is all about, I usually respond with

> *"The universal purpose of ham radio is to have fun messing around with radios."*

Sure, ham radio has a serious side involving public service communications in times of disaster or other events. But for the most part, ham radio activity is driven by the desire to have fun with radios.

I've been enjoying ham radio and writing about it ever since I received my first FCC license. Although I've used the ham bands from 160m to 10cm (10 GHz), for some reason I always find the VHF frequencies to be the most interesting. (And when I say VHF, I usually just mean any operating frequency above 50 MHz. I tend to mix VHF and UHF together into the same general category.) This is mainly due to the interplay between mountains, Height Above Average Terrain (HAAT) and VHF propagation. VHF has lots of spectrum, lots of modes and lots of fun challenges.

The first few articles in this book are popular tutorials from my website, followed by more traditional blog posts. I included some of my best posts about Summits On The Air (SOTA), but tried to not overdo it. There are many more SOTA trip reports on my blog, so check those out if you are interested. I've updated some of the posts that were out of date and added a few new ones to help round out the book. Also, I edited everything for improved grammar and clarity.

Of course, I am interested in your questions and feedback. Stop by my website at www.k0nr.com to see my latest work.

73, Bob KØNR
bob@k0nr.com

2. Frequency, Band and Wavelength

The table below is a quick guide to help sort out the terminology radio hams use for describing their operating frequency. There are three large frequency ranges that we often refer to: HF, VHF and UHF, defined by the first and second columns of the table below. The amateur radio allocations within those broad ranges are listed in the third column. The wavelength of the band is listed and the specific frequency range allowed for amateur use (US Allocations).

Range	Frequency	Amateur Allocation (band: frequency)
High Frequency (HF)	3 to 30 MHz	80 m through 10 m
Very High Frequency (VHF)	30 to 300 MHz	6 m: 50 to 54 MHz 2 m: 144 to 148 MHz 1.25 m: 222 to 225 MHz
Ultra High Frequency (UHF)	300 to 3000 MHz	70 cm: 420 MHz to 450 MHz 33 cm: 902 to 928 MHz 23 cm: 1240 to 1300 MHz 13 cm: 2300 to 2450 MHz

PART I
VHF OPERATING

This section provides a basic introduction to the VHF/UHF bands and operating practices.

PART 1
VHF OPERATING

3. VHF FM Operating Guide

This guide is intended to assist new amateur radio operators in figuring out what VHF/UHF FM and repeater operation is all about. This is written based on my own personal experience operating in the Midwestern and Western states of the US. Some references are specific to my state of Colorado. You may find that certain operating practices are different in your area. I encourage all new hams to *do a lot of listening* when they first get their equipment. Try and figure out what the standard operating norms are in your location. And make sure to try to emulate the best quality operating practices and avoid the poor practices that are often found on the ham bands. (Note that "old timers" are often just as guilty of sloppy operating.)

REPEATER OPERATING

Repeater operation is a little bit different than other forms of amateur radio communication. Usually the signals are strong and clear so some of the practices used on the other bands are not necessary on VHF FM.

CHECKING INTO THE REPEATER

If you are out mobile (or just hanging around the ham shack) and want to talk to someone, flip to the repeater frequency and see if anyone is around. It's not usually necessary to give a long call on VHF FM. Most stations indicate their availability for a call by transmitting and saying "KØABC monitoring." Sometimes the term "monitoring" is interpreted as "I am here if anyone wants to talk to me" as opposed to "calling any station." You might try saying "This is KØABC, anyone around?" If you have a specific request, such as a signal report, traffic or weather information, say so. Someone will be much more likely to jump in on a specific request for help.

IDENTIFICATION

FCC regulations require an amateur operator to identify at the end of a series of transmissions and at least every ten minutes during a series of transmissions. It is considered good practice to identify your station when you first come on the air (even though the FCC doesn't require it). This lets other stations know who you are right from the start.

BREAKING

Breaking in on a conversation is just that, interrupting someone else when they are talking. A repeater is a shared resource, so we should expect that other people may have a need to use the machine while we are on the air. No one has exclusive rights to the frequency and repeater.

The best way to break into a conversation is to simply give your call sign in between transmissions. Some operators use the word "break" which is somewhat controversial. In some circles it is perfectly acceptable, while in others it is reserved for emergency communications. Usually, if you have time to squeeze the word "break" in, you can just as well squeeze in your call. Using your call sign instantly identifies yourself to the stations using the repeater and often results in a "Go ahead, Joe" instead of "Is there someone trying to break in?"

During Amateur Radio Emergency Service (ARES) operations, our local ARES organization reserves the word "break" for emergencies. When an emergency net gets busy, multiple stations may need to break in. If three stations end up saying "break", the net control station has a hard time sorting things out. If call signs are used, the net control can make sure all stations are acknowledged and their traffic handled. Reserving the word "break" for emergencies is a good habit to adopt for general repeater use.

Of course, if operators using the repeater make it a point to pause between transmissions, it makes breaking in that much easier and more orderly. Most repeaters have a "courtesy beep" of some sort which forces operators to wait for a short time between transmissions. If they don't wait, they run the risk of timing out the machine and having it shut down on them.

Most FM repeater operating can be done with a minimum of slang and jargon. Plain language works quite well. Try to avoid the overuse of Q signals, phonetics, etc. such as "Roger, roger, Joe, I QSL your QSL and thanks for the QSO from your QTH, roger, roger."

PURPOSE OF A REPEATER

In the strictest sense, the purpose of a repeater is to extend the range of handheld and mobile stations. Base stations with reasonable antennas often don't need the extended range afforded by a repeater. However, it is common practice for all types of stations to make use of repeaters. We need to think of the repeater as a valuable resource (especially wide coverage repeaters). If you are talking with someone who is within simplex range, go ahead and switch over to a simplex frequency and free the repeater up for other people to use. This is especially true if you are going to ragchew for quite a while.

A repeater is a shared resource. It takes a considerable amount of time and money to keep a repeater up and running and there are many people who use the machine. Thus, courtesy is the order of the day. Yield the frequency to someone who needs it. Also, try not to interrupt another conversation on the repeater needlessly. Cooperate with other operators to keep the repeater operation fun and useful. Think about the content of your repeater conversations since many other radio operators and non-licensed individuals may be monitoring the frequency. If you've run out of useful things to say, it's probably time to sign clear! The most important two things to remember about repeater operation (for that matter, any amateur radio operation) are common sense and courtesy.

SIMPLEX OPERATING

Repeaters are very powerful and useful tools for amateur communications. They can dramatically extend the range of a handheld or mobile transceiver and they offer extended capabilities such as autopatch, weather station access, internet linking, etc. All of the emphasis on the utility of repeaters can lead the new ham to conclude that repeaters are *the only thing* worth using on the VHF/UHF bands. But this is not true!

Simplex operation is direct station-to-station radio communication without the use of an intermediate relay station (i.e., repeater). The range is much more dependent on antenna type, antenna height and power output than repeater operation but significant distance can be covered. Two mobiles having a QSO on flat land can typically cover 15 to 20 miles. If one station is on a hill, the range can be much greater. If you happen to be standing on top of Pikes Peak (at 14,115 feet about sea level), you can contact someone *several hundred miles away*. See Chapter 17: VHF Distance From Pikes Peak. Which leads me to my next point: simplex operation can be fun. It is a challenge to see how far your signal can go (yes, *without* some powerful repeater extending your range).

So don't forget about simplex. Some of the best contacts I ever made were on simplex. One of the reasons is the challenge of seeing how far your signal will go. The other factor is that there are fewer people listening on simplex. There is less of the "party line atmosphere" that can exist on a repeater where everyone within repeater range will hear every word of every station. Also, you are not tying up the repeater. Often, this leads to a longer and more meaningful discussion between you and the operator on the other end.

For the best performance on simplex, use SSB (single-sideband) or CW (continuous wave, Morse Code). You will find that all of the really serious VHF weak-signal operators use SSB and CW. The focus of this chapter is on the FM mode, so I won't go into great detail here. But I must point out that for getting the most out of simplex operation, you'll want to check out the non-FM modes. See Chapter 4: Getting Started on 2m SSB.

> Give Simplex a Try.

MAKING CONTACT

Sometimes new hams find that no one answers their call. So they try again. Again, there is no answer. After a few days of trying this, they figure that the local ham operators are ignoring them, probably because they have a "new" call sign. Yeah, they've heard about the oldtimers being rude to the newcomers, so they assume that's happening to them. Well, they may be right, but probably not. Keep in mind that hams that have been on the air for a long time often view their radio as a means of accomplishing certain things...like talking to their buddies or checking into their favorite nets. The thrill of meeting new, unknown people may be gone. So they become reluctant to grab the microphone to answer every new call. But they aren't out to get you.

My advice? Chill out. Don't put a chip on your shoulder if people don't flock to respond to your call. You've just walked into an electronic cocktail party and sometimes it is difficult to strike up a conversation. Listen for when someone else checks in on the frequency and call them instead. Make sure you have something interesting to say besides "The rig here is a BelchFire 602 running 1kW into a melted rubber duck antenna." Try to find people that have interests in common with you. Another strategy is to attend one of the local ham clubs and get to know a few people. Volunteer for club activities, attend Field Day and, in general, be willing to help out. Once you meet people in person, the on-the-air contacts are more frequent and meaningful.

PROGRAMMING THE TRANSCEIVER

It is important that you get your transceiver programmed correctly so that you can access the desired repeater or simplex frequency. For repeater operation, you need to set the *Frequency*, *Offset* and *Tone* (FOT). *Frequency* is the transmit frequency of the repeater and your receive frequency. *Offset* is the transmit frequency offset and *Tone* is the CTCSS tone often required to access a repeater. For a more complete discussion of FOT, see Chapter 11: FOT: Frequency, Offset and Tone.

TRANSMITTER TESTING

Try to keep test transmissions off the air by using a good dummy load. Be aware that even a slight amount of radio energy leaking from a dummy load may activate a sensitive repeater. Use a transmit frequency other than the repeater input even when testing with a dummy load. Some tests must be performed on the air, such as tuning an antenna for best match. Use a repeater frequency that is not used in your vicinity or a not-too-popular simplex frequency.

If your intent is to key the repeater to see how strong its signal is, don't kerchunk it. Instead, transmit and say "KØABC testing." Remember, unidentified transmissions are illegal and annoying. Also, most repeaters have special circuitry which detects kerchunking and sends out a powerful signal that causes the offending radio to burst into flames. Just kidding.

INTERFERENCE

Most repeaters in Colorado are blessed with very few malicious interference problems. If you do hear this sort of thing (jamming, foul language, etc.), NEVER acknowledge it on the air. Giving the person some kind of response only encourages the behavior. Also keep in mind that interference may not be intentional. Most operators have at one time or another accidentally keyed the mike or have mistakenly left the receiver volume turned down and transmitted on top of someone. Assume good intentions until proven otherwise.

PROPAGATION

Hams usually consider both 2m and 70 cm to exhibit line-of-sight propagation. This means that the signal travels to the optical horizon (and perhaps a little farther). Increased Height Above Average Terrain (HAAT) increases the distance to the horizon and propagation distance. This is why repeaters are located on the tops of mountains or tall buildings.

The front range of Colorado enjoys excellent VHF & UHF repeater coverage due to the close proximity of the mountains to the flat plains where much of the population is concentrated. Propagation characteristics of 144 MHz and 440 MHz are similar, with 440 MHz more susceptible to the shadowing effects of hills and other obstructions.

On some occasions, VHF and UHF propagation enhancement occurs and signals propagate significantly further than line-of-sight. These propagation modes include tropospheric ducting, sporadic E E and meteor scatter. Most of the long-distance VHF/UHF work is done using single-sideband or CW on the low end of the bands. This is a whole 'nuther aspect of VHF operating and is a ton of fun but not the focus of this chapter.

SIGNAL REPORTS

FM signal reports are often given in terms of receiver quieting. A strong signal will fully quiet an FM receiver, while a weak one will be quite noisy. A "full quieting" report is given to a signal which exhibits no background noise or hiss. Signal reports are often given in terms of "percent quieting" to give the transmitting station a better idea of the signal quality.

Most VHF FM radios have a meter or bar graph that indicates received signal strength, commonly referred to as an S meter. A true S meter will be calibrated to read out signal strength from S-1 to S-9, with S-9 being the strongest signal. However, many S meters provide only a relative indication of signal strength using a bar graph. It is common for people to give S-9 reports for full-scale meter readings and estimate a lower signal report for smaller signals (say, S-3 or S-5). The point is that these S meters are not very accurate but may still provide useful information.

Remember that when using a repeater there are two communication paths at work — the path from the transmitting station to the repeater and the path from the repeater to the receiving station. Either one of these paths can exhibit noise due to a weak signal. If the receiving station has a strong S-meter indication but the transmitting station sounds noisy, the transmitting station is probably weak into the repeater. Remember that the signal strength indicated by your S meter is due to the repeater and not the transmitting station.

A frequency-modulated transmitter used on the VHF/UHF amateur bands should be set for a maximum frequency deviation of 5 kHz with full modulation. Unlike SSB transmitters, the signal strength of an FM signal is independent of modulation level. That is, a dead carrier produces just as much power as a fully-modulated signal. Excessive modulation of an FM transmitter does not improve the reception of the signal and often degrades it. On the other hand, inadequate FM deviation causes weak received audio. The level of audio heard on the receive end is relatively independent of received signal strength. This means that changing transmitter power does not affect the loudness of the audio at the receive end.

PHONETICS

The use of phonetics is not usually required due to the clear audio normally associated with frequency modulation. Still, sometimes it is difficult to tell the difference between similar sounding letters such as "P" and "B". Under such conditions, use the standard ITU phonetics to maintain clarity. Many nets specifically request the use of standard phonetics to make it easier on the net control station. See Chapter 28: Go Ahead and Use Phonetics on 2m FM.

Q SIGNALS

Q signals are most useful when using Morse Code and are normally not required on phone. However, they are part of the amateur radio culture and are used on the air. The following is a short list of common Q signals, as used on VHF.

Q Signal	As a statement	As a question
QSY	I am changing frequency (to)	Can you change frequency (to)?
QSL	I acknowledge receipt (of a message)	Do you acknowledge receipt (of a message)?
QSO	I can communicate with (station) More informally: refers to a radio contact, as in "I just had a QSO with Fred this morning"	Can you communciate with (station)?
QTH	My location is _____	What is your location?
QRM	I am being interfered with	Are you being interfered with?

EMERGENCIES

Radio amateurs have a long history of helping out when during emergencies. These emergencies tend to fall into two categories: 1) Disaster situations when the ARES or RACES organizations are activated and 2) Short-term emergencies that a single radio op happens upon. Radio amateurs are urged to participate in their local ARES or RACES organizations to be fully prepared for the first type of emergency.

The second category of emergencies will be discussed further here. A common scenario is needing to call the proper authorities to report a traffic accident or other form of event that is causing property damage or injury. First, be aware that in areas with good mobile phone coverage, it will be more effective to report an emergency situation via telephone. Using a mobile phone provides a direct, full-duplex connection to the emergency (911) dispatch center. It also provides a mechanism for the 911 dispatcher to call you back.

If you need to use amateur radio to request help, you'll probably be talking to another radio ham that will relay your message to the proper authorities. Some things that you need to think about in an emergency:

- Where are you?
- What is the nearest mile marker, intersection or landmark?
- What type of assistance do you require?
- Are there any injuries? (The authorities want to know whether to dispatch medical or not.)

VHF/UHF BAND PLANS

The general band plan for the amateur 2-Meter band as proposed by the ARRL VHF-UHF Advisory Committee is:

144.00-144.05 EME (CW)
144.05-144.06 Propagation beacons
144.06-144.10 General CW and weak signals
144.10-144.20 EME and weak signal SSB
144.20 National SSB calling frequency
144.20-144.30 General SSB operation
144.30-144.50 OSCAR sub-band
144.50-144.60 Linear translator inputs
144.60-144.90 FM repeater inputs
144.90-145.10 Weak signal and FM simplex (includes packet radio)
145.10-145.20 Linear translator outputs
145.10-145.50 FM repeater outputs
145.50-145.80 Misc. and experimental modes
145.80-146.00 OSCAR sub-band
146.01-146.37 Repeater inputs
146.40-146.58 FM simplex
146.61-147.39 Repeater outputs
147.42-147.57 FM simplex
147.60-147.99 Repeater inputs

Note that the 2-Meter band includes both FM and weak-signal (SSB/CW) operation. FM users of the band should not stray into the weak-signal or OSCAR (satellite) sub-bands.

Specific 2m FM frequencies used in Colorado have 15 kHz spacing. All repeater output frequencies below 147 MHz have the repeater input -600 kHz relative to the repeater output. All repeater output frequencies above 147 MHz have repeater inputs +600 kHz relative in the repeater input. Most modern 2-Meter transceivers are tuned by selecting the repeater output frequency and then selecting the transmit offset (+ or – 600 kHz).

Standard simplex frequencies in Colorado are:

146.415, 146.430, 146.445, 146.460, 146.475, 146.490, 146.505, 146.520, 146.535, 146.550, 146.565, 146.580, 146.595, 147.420, 147.435, 147.450, 147.465, 147.480, 147.495, 147.510, 147.525, 147.540, 147.555, 147.570, 147.585 MHz

> **The National Simplex Calling Frequency is 146.52 MHz.**

The 440-MHz band plan as proposed by the ARRL VHF-UHF Advisory Committee is:

420.00-426.00 ATV repeater or simplex, control links and experimental
426.00-432.00 ATV simplex
432.00-432.07 EME
432.07-432.08 Propagation beacons
432.08-432.10 Weak signal CW
432.100 SSB calling frequency
432.10-432.125 Mixed mode and weak signal
432.125-432.175 OSCAR inputs
432.175-433.00 Mixed mode and weak signal
433.00-435.00 Auxiliary/repeater links
435.00-438.00 Satellite only
438.00-444.00 ATV repeater input and repeater links
442.00-445.00 Repeater inputs and outputs
445.00-447.00 Auxiliary and control links, repeaters and simplex
446.00 National Simplex (Calling) Frequency
447.00-450.00 Repeater inputs and outputs

Standard 440 MHz FM frequencies are spaced 25 kHz apart. In Colorado, the transmit offset on 440 MHz repeaters is -5 MHz (that is, the repeater input frequency is 5 MHz below the output frequency.)

VHF/UHF band plans are managed regionally, so if you are not in Colorado check with your local repeater coordinating body. For more information on frequency selection, see the articles in the Reference section.

REPEATER DIRECTORY

The American Radio Relay League (ARRL) publishes a repeater directory every year. This directory lists the VHF/UHF repeater frequencies for the entire United States, Canada, the Caribbean, Central and South America and the Pacific islands under U.S. jurisdiction. This directory is highly recommended for radio amateurs who travel outside their local area. Contact the ARRL at http://www.arrl.org/.

Recently, a number of online repeater directories have popped up on the web with some of them having smartphone apps. The Repeater-Book.com website seems to be up to date and easy to use.

ANTENNAS FOR HANDHELD RADIOS

Almost everyone routinely uses a "rubber duck" antenna for handheld transceivers (HT). This antenna is essentially a quarter-wave which is shrunk down to about one-fourth of its usual length. Think of this antenna as a leaky dummy load, because its effectiveness is not much better than that. Its short length and the lack of a ground plane (which is required for a quarter-wave antenna) makes its performance quite poor. Only the high sensitivity of FM repeaters make handheld radios with rubber duck antennas so useful. Of course, the small size of the antenna is very convenient for portable use, which is why it is so commonly used.

For hiking, public service events, and other activities where radio range is important, a longer antenna is very helpful. Although there are full-size quarter-wave and 5/8-wave models available, experience has shown that one of the most effective handheld antennas is the end-fed half-wave. The telescoping half-wave antenna with a BNC or SMA connector on the end is available from several manufacturers. This type of antenna can easily make the difference between having an unreadable signal and being full-quieting into the repeater.

When a good antenna is attached to an HT, the receiver may exhibit problems due to the much stronger signals present. Strong signals may come blasting through the receiver and interfere with the desired signal. Radio amateurs usually refer to this as "intermod," short for "intermodulation." In reality, intermodulation has a specific technical definition that describes only some of these noise and interference problems. Independent of the name, the end result is that the HT receiver is overloaded by these strong signals. One solution to the problem is to use an external filter to block out signals outside the ham band. This type of filter will also block police, fire, weather and other non-ham signals, too.

References

What Frequency Do I Use on 2 Meters?
http://www.hamradioschool.com/what-frequency-do-i-use-on-2-meters/

What Frequency Do I Use on 70 Centimeters?
http://www.hamradioschool.com/what-frequency-do-i-use-on-70-centimeters/

4. Getting Started on 2m SSB

In the past decade, a new breed of amateur radio transceiver has hit the marketplace — radios that cover from HF through VHF/UHF frequencies. These radios include the ICOM IC-7100, the Yaesu FT-857 and the Yaesu FT-991A. This is not an exhaustive list since there are new radios being introduced every year with additional capability. These radios include "all-mode capability" which means that they can operate FM (Frequency Modulation), CW (Continuous Wave) and SSB (Single Sideband) on the VHF bands. Clearly, FM is the most commonly used mode on VHF and UHF but having SSB opens up a whole new range of operating fun.

The Yaesu FT-991A is an example of a transceiver that covers HF, VHF and UHF.

Why SSB?

FM is the most popular mode primarily due to the wide availability of FM repeaters. These repeaters extend the operating range on VHF and enable low power handheld transceivers to communicate over 100 miles. FM is also used on simplex to make contacts directly without repeaters. The main disadvantage of FM is relatively poor performance when signals are weak, which is where SSB really shines. A weak FM signal can disappear completely into the noise while a comparable SSB signal is still quite readable. How big of a difference does this really make? Perhaps 10 dB or more, which corresponds to one or two S-units. Put a different way, using SSB instead of FM can be equivalent to having a beam antenna with 10 dB of gain, just by changing modulation types. So this is a big deal and radio amateurs interested in serious VHF work have naturally chosen SSB as the preferred voice mode. (You will also hear them using Morse code or CW transmissions, which is even more efficient that SSB.)

Just as an example of what is possible on SSB, during one VHF contest I was operating portable on Garden of the Gods Road in Colorado Springs. I had just dismantled my 2-meter Yagi antenna and was listening to 2-meter SSB on a short mobile whip antenna. Suddenly, I heard WA7KYM in Cheyenne, Wyoming calling CQ from about 160 miles away. I figured that with my puny little antenna and only 10 watts of power, there was no way he was going to hear me. But, what the heck, it was a contest and it would be more points so I gave him a call. To my surprise, WA7KYM heard me and we made the contact without much signal strength to spare. Now, to be accurate, this contact has more to do with WA7KYM's "big gun" station (linear amplifier, low noise preamp and large antenna array) than it had to do with my 10 watts and a small whip. The key point here is that this contact would not have happened using FM and was only possible because of SSB.

When and Where to Operate

The SSB portion of the band runs from 144.100 MHz to 144.275 MHz and Upper Sideband (USB) is used. The 2-meter SSB calling frequency is 144.200 MHz, so that is the first place to look for activity or to call CQ. One of the realities of 2-meter SSB operation is that many times, no one is on the air. There is just not that much activity out there, compared to 2-meter FM. Some amateurs get discouraged, turn off the radio and miss the thrill of working distant stations during a band opening. To get started on 2-meter SSB, the trick is to get on the air at times when you know there will be activity— during VHF nets and VHF contests.

Here in Colorado, the local Rocky Mountain VHF Plus net is on Monday night at 8:00 PM local time on 144.220 MHz (USB). This net is centered in the Denver area but VHF enthusiasts check in from all around Colorado. It is common to have stations check in from the bordering states of Wyoming, Nebraska, Kansas, New Mexico or even Oklahoma.

To find nets in your area, search on the internet with your state or location and "amateur radio ssb vhf net."

VHF Contests

Think of VHF contests as "VHF activity weekend" since they are a great opportunity to just get on the air and work most of the local 2-meter SSB enthusiasts. The main contests are the ARRL June VHF Contest, the ARRL January VHF Contest, the ARRL September VHF Contest and the CQ Worldwide VHF Contest in July. To learn a bit about VHF contests, take a look at Chapter 7: How to Work a VHF Contest.

Equipment

The required equipment for getting started on 2-meter SSB is pretty basic – a transceiver capable of 2-meter SSB and a 2-meter antenna. If you own one of the rigs mentioned previously then you are probably ready to go. The 2-meter antenna you already have is probably vertically polarized since that is what we use for FM, both mobile and base stations. The quarter-wave and 5/8-wave antennas that are commonly used for 2-meter mobile work are vertically polarized. Most omni-directional base station antennas such as those made by

Cushcraft, Diamond, Comet, etc. are vertical, too. These antennas will work for SSB but most of the active 2-meter SSB stations use horizontally-polarized antennas. Vertically-polarized stations can work horizontally-polarized stations but there will be a substantial signal loss (perhaps 15 to 20 dB). If vertical is all you have, then give it a try. If you can get a horizontal antenna in place, your results will be much better.

The most common horizontally-polarized antenna on 2 meters is a Yagi mounted so that its elements are parallel to the ground. There are a variety of horizontally-polarized, omni-directional mobile antennas, such as the HO antenna made by M^2 Antenna Systems.

Get on the Air

This information is intended to get you started on your way to operating 2m on the SSB portion of the band. You will learn more as you get into it and you will find that most of the people hanging out down on sideband are friendly, knowledgeable and helpful. They are always happy to hear new call signs on the band.

5. Six Meters: The Magic Band

The 6-meter band (50 to 54 MHz), often referred to as the *Magic Band*, is a fun band to operate. Most of the time, the propagation provides relatively local radio coverage, sometimes referred to as "line of sight." However, when the band opens up (usually due to sporadic-E propagation), 6 meters has skywave propagation and acts like an HF band. It is this unpredictable nature of the propagation that somehow delights us and causes us to label it the *Magic Band*.

The ICOM IC-7300 is an example of an HF transceiver that includes the 6-meter band.

Modulation Type

SSB is commonly used for phone contacts, especially for taking advantage of sporadic-E openings. FM is used for both simplex and repeater operation, although there are much fewer 6-meter repeaters (as compared to 2 meters). I've worked sporadic-E openings using FM but there's almost always more stations on SSB. CW is also quite popular.

The new digital mode, FT8, is seeing a lot of use on the 6-meter band. FT8 enables radio contacts to be completed even when signals are very weak. This improves our likelihood of making long distance 6-meter contacts because FT8 gets through to allow a basic QSO to be completed. FT8 is optimized for exchanging the minimum information (call sign, signal report or grid) to complete a contact, so it is not used for conversations.

6-meter Band Plan

Frequency (MHz)	Usage
50.0 to 50.1	CW only
50.060 to 50.080	Beacons
50.100 to 50.125	DX Window
50.125	SSB Calling Frequency
50.125 to 50.30	SSB
50.280	MSK144 Meteor Scatter
50.313	FT8
52.0 to 54.0	FM Repeaters and simplex
52.525	FM Calling Frequency

Equipment

In recent years, the ham equipment manufacturers have been including the 6-meter band on many of their HF transceivers. Stretching the design of the HF rig up to 54 MHz is not too difficult and it offers a nice "bonus band" for hams wanting to do some VHF work. The 6-meter band is also included on the "do everything" radios (e.g., Yaesu FT-991A or Icom IC-7100) that cover HF through 70 cm. Single-band 6-meter radios are not common today but you may find one on the used market.

For an antenna, you can use something as simple as a half-wave dipole, about 9 feet long, strung horizontally between two supports. Alternatively, you can use other wire antennas such as an end-fed half wave. Some hams use a quarter-wave vertical on 6 meters. For local communication, you need to be aware of the antenna and radiated signal polarization. For local communications, the signals will be much stronger if both stations have the same antenna polarity. For skywave propagation, antenna polarization doesn't matter. Most serious operators on 6m are going to opt for a Yagi antenna, horizontally mounted with perhaps 3 to 6 elements.

Propagation

Propagation on the 6-meter band is a lot like the 2-meter band, most of the time. Signals propagate nicely to the radio horizon and a bit beyond. Tropospheric scatter can extend the radio range well beyond the radio horizon and out to several hundred miles. This is not an exotic mode; it happens on a daily basis.

The 6-meter band often experiences skywave propagation via the E and F layers of the ionosphere. F-layer propagation occurs with high levels of solar activity, similar to the HF bands. In recent years, we haven't seen sufficient solar activity to cause much F-layer propagation on 6m. In that regard, the 6- meter propagation is a lot like the higher HF bands (e.g., 10 and 15 meters).

The E layer provides skywave propagation that occurs unpredictably, hence the name *sporadic-E propagation*. Sporadic E does not depend on high solar activity but does exhibit a seasonal variation with more frequent openings during the months of May to August, and to a lesser extent December and January. Sporadic-E clouds form randomly at a height of 50 to 100 miles, resulting in a single-hop distance of 800 to 1200 miles. Multiple-hop propagation also occurs extending the sporadic-E radio range to thousands of miles. Sporadic E can produce exciting but fickle operating conditions. During the summer months, the band can be very quiet with only local stations being heard followed by a really strong opening to other parts of the country. These openings can be very brief or may last for hours.

SSB Operating

The first place to look for SSB activity is to tune into the calling frequency on 50.125 MHz, using Upper Sideband (USB). You can listen there for anyone making a call and you can go ahead and call CQ on that frequency. If the band opens up for distant propagation, the activity usually moves upward in frequency, perhaps as high as 50.300 MHz. The DX window from 50.100 to 50.125 MHz is reserved for contacts with stations outside of the continental US. So don't work other stateside stations in the window as it is set aside for more distant contacts.

> **Calling frequencies on the 6-meter band are 50.125 MHz (SSB) and 52.525 MHz (FM).**

What About FM?

FM operation on 50 MHz can be via simplex or repeaters. Six-meter repeaters are not very common but if you are lucky to have some in your area you may find a community of enthusiastic 6-meter FM operators. For simplex operation, try the 6-meter FM Calling Frequency: 52.525 MHz.

The 6-meter band is a fun band, but one that can be fickle. This means you have to be patient and look for band openings to make distant contacts.

6. The Myth of VHF Line of Sight

When we teach our Technician License class, we normally differentiate between HF and VHF propagation by saying that HF often exhibits skywave propagation but VHF is normally line-of-sight. For the beginner to ham radio, this is a reasonable model for understanding the basics of radio propagation. As George E. P. Box said, "All models are wrong, but some are useful."

In recent years, I've come to realize the limitations of this model and how it causes radio hams to miss out on what's possible on the VHF and higher bands.

Exotic Propagation Modes

First, let me acknowledge and set aside some of the more exotic propagation modes used on the VHF and higher bands. **Sporadic-E propagation** allows long distance communication by refracting signals off the E layer of the ionosphere. This is very common on the 6-meter band and less so on the 2-meter band. I like to think of this as the VHF bands trying to imitate HF. **Tropospheric ducting** supports long-distance VHF communication when ducts form between air masses of different temperatures and humidities. **Auroral propagation** reflects the radio signal off the auroral ionization that sometimes occurs in the polar regions. **Meteor scatter** reflects signals off the ionizing trail of meteors entering the earth's atmosphere. **Earth-Moon-Earth (EME)** operation bounces VHF and UHF signals off the moon to communicate with other locations on earth. These are all interesting and useful propagation mechanisms for VHF and higher but not the focus of this article.

Line-of-Sight Model

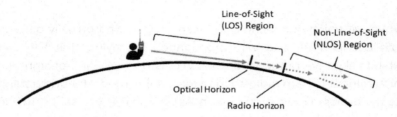

Now let's take a look at more "normal" VHF propagation that occurs on a daily basis, starting with the simple line-of-sight propagation model. The usual description of line-of-sight VHF is that the radio waves travel a bit further than the optical horizon (say 15% more). Let's refer to this as the Line-of-Sight (LOS) region where signals are usually direct and strong. What is often overlooked is that beyond the radio horizon, these signals continue to propagate but with reduced signal level. Let's call this the Non-Line-of-Sight (NLOS) region. The key point is that the radio waves do not abruptly stop at the edge of the LOS region...they keep going into the NLOS region but with reduced signal strength. Now I will admit that this is still a rather simplistic model. Perhaps too simplistic. I'm sure we could use computer modeling to be more descriptive and precise, but this model will be good enough for this article. *All models are wrong, but some are useful.*

Another scenario occurs when mountains (or other geographic features) block the path of the radio signal. We encounter this scenario while doing mountaintop operating with other mountains in the area blocking our radio signals. A simple line-of-sight model would say that a VHF radio signal will be blocked by a mountain and will only travel so far.

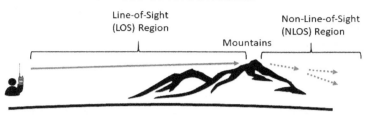

Effect of Mountains

Again, we show the LOS region where the signal has a direct shot and an NLOS region where the radio waves keep on going with lower signal strength. The actual propagation path to stations in the NLOS can be very complex. The signal may be propagated by knife-edge refraction as the radio wave encounters the top edge of a mountain. The signal may find its way by reflecting off the various mountain slopes and bouncing its way through to the other side. In a practical situation, we may have no idea how the signal gets through and we may not be able to predict how strong it will be in the NLOS zone. Again, this model of LOS and NLOS is pretty simple but it can help us understand what is happening.

WORKING the LOS and NLOS Regions

Let's apply the model for Summits On The Air (SOTA) VHF activations. If we are only interested in working the LOS region, we won't need much of a radio. Even a handheld transceiver with a rubber duck antenna can probably make contacts in the LOS region. It's still worth upgrading the rubber duck antenna to something that actually radiates to improve our signal (such as a half-wave antenna). This will help us make radio contacts in the NLOS region, but but our success may be limited.

To improve our results in the NLOS zone, we need to increase our signal strength. We are working on the margin, so every additional dB can make enough difference to go from "no contact" to "in the log." Think about another radio operator sitting in the NLOS zone but not quite able to hear your signal. Your signal is just a bit too weak and is just below the noise floor of the other operator's receiver. Now imagine that you improve your signal strength by 3 dB, which is just enough to get above the noise and be a readable signal. You've just gone from "no contact" to "just readable" with only a few dB of improvement.

What can we do to improve our signal levels? The first thing to try is improving the antenna, which helps you on both transmit and receive. I've already mentioned the need to ditch the rubber duck on your HT. My measurements indicate that a half-wave vertical is about 8 dB better than a typical rubber duck. This is only an estimate...performance of rubber duck antennas vary greatly. A small Yagi antenna (3-element Arrow II Yagi) can add another 6-dB improvement over the half-wave antenna, which means the Yagi has about a 14 dB advantage over a rubber duck.

On the other hand, if you believe that your VHF radio is only Line-of-Sight, then there is no reason to work on increasing its signal level. The radio wave is going to travel to anything within the radio horizon and then it will magically stop. This is the myth that we need to break.

More Power

When doing SOTA activations, I noticed that I was able to hear some stations quite well but they were having trouble hearing me. Now why would this be? Over time, I started to realize that they were typically home or mobile stations running 40 or 50 watts of output. This created an imbalance between the radiated signal from my 5 W handheld and their 50 W radio. In decibels, this difference is 10 dB. Within the LOS region, this probably is not going to matter because signals are strong anyway. But when trying to make more distant radio contracts into the NLOS zone, it definitely makes a difference. So I traded my HT for a mini-mobile transceiver running 25 W. See the complete story in Chapter 51: More Power For VHF SOTA.

Weak-Signal VHF

Of course, this is nothing new for serious weak-signal VHF enthusiasts. They operate in the NLOS region all of the time, squeezing out long-distance QSOs using CW, SSB and the WSJT modes. They generally use large directional antennas, low noise preamps and RF power amplifiers to improve their station's performance. They know that a dB here and a dB there adds up to bigger signals, longer distances and more radio contacts.

Now you might think that FM behaves differently, because of the *threshold effect*. When FM signals get weak, they fade into the noise quickly...a rather steep dropoff compared to SSB which fades linearly. FM has poor weak-signal performance AND it fades quickly with decreasing signal strength. This is why it is not the favored modulation for serious VHF work. But the same principle applies: if we boost our signal strength by a few dB, it can make the difference between making the radio contact or not.

7. How to Work A VHF Contest

This is a brief introduction into how to operate during a VHF contest. The main contests, roughly in order of popularity, are the ARRL June VHF Contest, the ARRL January VHF Contest, the ARRL September VHF Contest and the CQ Worldwide VHF Contest in July.

The ICOM IC-9700 transceiver covers the 2m, 70cm and 23cm bands.

I prefer to think of these "contests" as "activity weekends" because the word "contest" often makes people think of the fast-paced, chaotic, band-crushing experience of HF contests. VHF contests usually have a much different feel. The problem with the VHF bands is that they are often underutilized. You put out a call on simplex and nobody is there. Dead silence. But on VHF contest weekend, you are sure someone is going to be on the air, so the event tends to increase the activity, bringing people out of the woodwork. A VHF contest is more like a friendly reunion of local VHF enthusiasts.

(Sometimes a VHF contest can get pretty intense, especially if there is a significant band opening on 6 meters. Then things start to sound like the HF bands with signals coming in from across the country.)

Frequencies

Frequencies above 50 MHz (6 meters and higher) are used during the contest. Most of the operation will be on 6 meters and 2 meters, less on the higher bands. Most of the operation will be on the SSB portion of the band, so if you have an all-mode VHF rig, you'll want to use it.

SSB Calling Frequencies	
6 meters	50.125 MHz
2 meters	144.200 MHz
1.25 meters	222.100 MHz
70 centimeters	432.100 MHz

As these frequencies become busy, move up the band.

Most SSB operation on VHF is done using *horizontal* antenna polarization. A Yagi or dipole antenna with radiating elements parallel to the ground produces a horizontally-polarized signal. A vertical antenna, commonly used for FM, produces a vertically-polarized signal. Working a station with opposite antenna polarity causes a substantial signal loss, so it is best to maintain the same polarity. For serious SSB operators, this means horizontal polarization. Kent Britain WA5VJB designed a series of homebrew Yagi antennas, that are cheap and easy to construct.

CW (Morse Code) is used on the weak-signal VHF bands, often intermingled with SSB operation. It is fairly common to have a station switch from SSB to CW when signals are very weak, since CW will get through at lower signal levels. You don't need to be able to work CW to enjoy a VHF contest but it does have advantages.

If you only have FM gear, you will be at a disadvantage but you may still be able to work a bunch of stations. Starting in 2016 the ARRL contests allow the use of the 2-meter FM calling frequency, 146.52 MHz. Don't monopolize this frequency. If it gets busy, move off to any of the other standard simplex frequencies. Never use a repeater for contest contacts.

FM Calling Frequencies

6 meters	52.525 MHz
2 meters	146.52 MHz
1.25 meters	223.5 MHz
70 centimeters	446.00 MHz

Operating Categories

The VHF contests have quite a few operating categories to choose from, depending on the number of operators at a particular station, the power level and other factors.

Feel free to read the rules to understand what is available, but for beginners these are the most relevant categories for the ARRL contests (CQ Worldwide VHF Contest is a little different):

- **Single Operator Low Power** – this is the "standard" single operator category, without running a big amplifier. Power limits are 200 W PEP on 50/144 MHz; 100 W PEP on 222/432 MHz; 50 W PEP on 902 MHz and higher.
- **Single Operator 3 Band** – this category allows use of the 6m, 2m & 70cm bands only; 100 W or less on 6m/2m, 50 W or less on 70cm.
- **Single Operator Portable** – this is the QRP category, 10 W PEP or less, portable power source, portable antennas, cannot use a permanent station location.
- **Single Operator FM Only** – this is a new category for FM-only operation, using the bands 6 m through 70 cm, 100 W or less on all bands.

You might also try being a rover, which means operating from more than one grid during the contest. If you operate in a Rover category, you need to say the word "Rover" after your call sign so that other operators know that you will be operating from multiple grids:

- **Rover** – this is the "standard" rover category, 1 or 2 operators, use all bands above 50 MHz, 1500 W PEP maximum power

- **Limited Rover** – limited rover category, 1 or 2 operators, 6 m – 70 cm only; 200 W PEP or less on 6m/2m, 100 W PEP or less on 1.25m/70cm

As a beginner, you can probably just choose your category based on the equipment you have. Or, if you want to go mountaintopping, consider operating as Single Operator Portable. If you want to operate mobile from a number of grids, then choose one of the rover categories.

Making A Contact

OK, so you get on the right frequency and the right mode. Now what? You need to make a contact. An official contact requires that the two operators exchange call signs and grid locators. VHF grids are a system that divides the world up into rectangles that are 1 degree of latitude by 2 degrees of longitude. You can work each station once per band for contest credit (except for rovers, who get to work everyone again in a new grid).

For more information on grid locators, see the Chapter 12: VHF Grid Locators

All of Colorado Springs and Pueblo are in grid DM78.
All of greater Denver and Castle Rock are in DM79.
Longmont, Loveland and Fort Collins are in DN70.

If you are close to the edge of a grid, you will need a good map or a GPS receiver to determine your grid.

Plug it in, turn it on and work someone on VHF this weekend.

References

ARRL January VHF Contest – http://www.arrl.org/january-vhf

ARRL June VHF Contest – http://www.arrl.org/june-vhf

ARRL September VHF Contest – http://www.arrl.org/september-vhf

CQ Worldwide VHF Contest – https://www.cqww-vhf.com/

WA5VJB Cheap Yagi Antennas – http://www.wa5vjb.com/yagi-pdf/cheapyagi.pdf

8. So You Want To Be a Rover

This article is about getting started as a rover in VHF contests, with emphasis on operating in or around Colorado. Maybe you've thought about trying some rover operating but weren't sure how to get started, so this article may help. Rover operation can be as simple or sophisticated as you'd like it to be but it is always a lot of fun. Operating rover is often just a good excuse to load up the radio gear and head out on a ham radio road trip.

Step one in understanding rover operation is to read the contest rules carefully to understand the rules specific to rovers. The most popular VHF contests are listed in the Reference section. I won't cover the rules here except to say that the basic concept is that rovers accumulate points by moving from grid to grid, making contacts with stations multiple times. For example, you might operate from 4 to 6 different grids, working many of the same stations from each grid. This type of operation is extremely valuable here in Colorado since many of the Colorado (and Nebraska, Kansas) grids are not occupied by fixed VHF stations.

Equipment

The first question that comes up is "what equipment do I need?" Again, this will vary greatly depending on you how much time, money and energy you want to put into rover operation. I will focus on *getting started* and you can build from there.

A basic VHF portable or rover operation that is set up for Stop and Shoot.

The minimum capability for rover operation is 2-meter SSB capability with a horizontally polarized antenna. Obviously, this station needs to be portable so that you can move from grid to grid. It is also very useful to have FM capability with you...most of the contesters are usually on SSB but there are a growing number of FM stations to work. Don't rely on a vertical antenna to operate SSB, since all serious contest stations will be horizontally polarized. Using the wrong polarization will cost you 20 dB or more in signal loss. There are a number of omnidirectional, horizontally polarized antennas for mobile use such as the Halo, SQLOOP, HO, Big Wheel, etc. You can mount these antennas on your vehicle and operate while in motion. Another alternative is to use a small Yagi antenna (again, horizontally polarized). This will require a more advanced mounting scheme and may require you to "stop and shoot" to operate. Of course, a Yagi has a significant gain advantage over omnidirectional antennas. Even though the coax runs are short, use low loss line such as 9913 or LMR400. In recent years, various dealers are offering "Flex 9913" and "LMR400 Flex," which has low loss but with a stranded center conductor for good mechanical flexibility, which is great for rover operation.

There are two basic approaches to rover operation, *Run-and-Gun* and *Stop-and-Shoot*. Run-and-Gun means that you make contacts while in motion. This requires some careful thought as to how to point antennas (if they are directional), perform logging and not get into a traffic accident. Stop-and-Shoot means that you stop at an appropriate location, set up antennas and operate while stationary. Many stop-and-shoot operators also have the ability to at least listen on 2 meters while in motion, so they can stay in touch with contest activity. Stop-and-shoot stations need to be quick to set up and take down, so that minimum time is lost. Most beginner rovers will choose stop-and-shoot, as it is inherently less difficult.

After you have the basic 2-meter SSB station covered, you can expand your station via additional frequency bands and increased station performance.

Additional Frequency Bands

There is no "single band" contest category for rovers, so adding additional bands is an important way to improve your score, your competitiveness and your operating fun. Six meters (50 MHz) is probably the most important band to add to your rover operation, since this band is most likely to have propagation to distant grids. When 6 meters is open, you'll want to focus your operating on this band to quickly gain contacts and grids. Antennas for 6 meters are more of a challenge, since even a 3-element Yagi is quite large for mobile operation. Most operators tend to choose a small, omnidirectional antenna. Sometimes I've used a vertical on 6 meters, which does well on long distance propagation but suffers for local communication due to polarization loss.

The next band to add is probably 70 cm (432 MHz), due to its popularity, then 23 cm (1.2 GHz), 1.25 m (222 MHz), etc. Check out the contest rules to understand the increased points per contact for the higher bands. It is helpful to monitor multiple bands simultaneously, so that you don't miss an opportunity on one band while making contacts on another.

A pickup truck set up for Run and Gun operation with small Yagi antennas. (Photo: George Palecek, ABØYM)

Improved Station Performance

Improved antennas are probably the first place to look for increased station performance. Rovers tend to operate with marginal antennas (at least compared to the big gun fixed stations). Increased antenna gain or height benefits both your transmit and receive performance. Even a small Yagi has many dB of advantage over an omnidirectional antenna. Antenna height is important, so many rovers develop mast systems that allow their antennas to get high off the ground (20 to 30 feet), improving their station performance.

Mounting of antennas is a great challenge for rover operation, especially for run-and-gun operation. I won't go into detail here, but antenna mounting schemes that I have used include:

- Drive-on Mast Mount – The basic idea is to mount a mast on a flat board, which is held in place by driving your vehicle onto the board. You can make a mast system yourself with a 2×10 piece of lumber, a few pipe fittings and some TV-style mast.
- Luggage/Sports Rack Mounts -Use your existing luggage rack to mount antennas, or adapt one of the popular sports racks (Yakima, Thule, etc.). There are a variety of mounts available that

use 3/8-24 thread that can be really useful with 3/8-24 rods. Some of the best mounts are available at truck stops, targeted at the CB market.
- Magnetic Mounts -Lots of different mag mounts can be used, often with the 3/8 -24 thread.
- Bumper Mounts – ham, CB and whatever you can fabricate yourself.
- Hitch Mounts -Many rovers have developed schemes for mounting masts on the standard 2-inch square Class III hitch receiver found on many vehicles.

Many rovers use linear amplifiers to provide a boost in transmit power. Often VHF and UHF amplifiers have low noise preamps, which help on receive. Increased output power means more attention must be paid to supplying DC power to the station, which means use large wire. A 150 W 2m amplifier will draw about 20 A of 12 V DC, which is a lot of current. (Just 1/10th of an ohm wire resistance will drop the supply voltage by 2 V!) A 300 Watt amplifier draws about 48 amps.

For simple stations, you can just power the rig off your vehicle's 12 V battery. However, you will need to run some heavy gauge wire directly to the battery and you'll need to start the engine periodically so that you don't run the battery down. Trust me, this is easy to forget when 6 meters opens up and you are the target of a large pileup. Many rovers use a separate battery to power their station.

Location(s)

Once you figure out what equipment you will use, you need to decide where you will go. The run-and-gun approach will result in choosing an operating route, while the stop-and-shoot approach focuses on a set of operating locations. Either way, thinking through the route you will take is very important, since it affects how much operating time you will ultimately have in each grid.

Choice of operating location depends on three main things: grid rarity, accessibility and propagation to other stations. Grid rarity means that you want to operate from rare grids, ones that are not likely to be activated by other operators during the contest. For example, DM78, DM79 and DN70 along the front range of Colorado have large populations of hams and will be activated during the contest. Out on the eastern plains (DN80, DM89, DM88), there are typically no fixed stations active, so it is a great place to set up as a rover. By being in a rare grid, you offer a larger incentive for contesters to work a little harder to make contact with you. In addition, radio hams that are working towards the VUCC award will make it a point to work you.

Accessibility means that you need to be able to get to the location quickly. This is a serious concern since many locations in the mountains look attractive until you consider the time to get to there. Snow can block backcountry roads even in the June contest since it is still early in the summer season.

Operating locations should have good propagation to other stations. Good locations are higher than the surrounding terrain with an unobstructed view in all directions. Fundamentally, the point is to make contacts, so if you are on top of a mountain with awesome propagation paths to locations with no one to work, it doesn't contribute to your score. In Colorado, you'll want to consider your propagation path to the front range cities where most of the stations are located. A careful review of topographical maps will help you choose the best spots.

Maps, GPS and other aids

You'll need a reliable means of knowing what grid you are located in. This starts with a macro view of the state...where are the grid lines at a regional level? Chapter 12: VHF Grid Locators has a Colorado map with grid lines drawn on it. It is also useful to have a standard state road map (one with roads and some geographic info on it) with grid lines drawn on it. This will give you the big picture on where you want to operate as a rover, but you will generally need something more precise to determine your exact location. The obvious choice is a GPS receiver or smartphone app that can be set to display the grid locator directly. Another alternative is a good set of topographic maps.

A laptop computer is very handy for logging purposes. Make sure your contest logging software knows how to handle rover stations. It has to be able to track which grid you are in and know that you can work stations multiple times, from each grid. Normal logging programs will flag that as a duplicate. One popular logging program that can handle rover operation is RoverLog.

Some rovers still use paper logs, but you'll need to have a scheme for keeping track of who you've worked from each grid. Otherwise, after working 3 bands from 6 grids, you can't remember if you need a QSO with a particular station or not. Some rovers just audio record their entire effort and sort it out later.

FM

As mentioned earlier, the vast majority of contest operation is on SSB (and CW). FM can be used in contests but check the rules to see if 146.52 MHz is allowed. My rover vehicle has a 2m/70cm rig installed permanently, so I usually monitor the standard FM simplex frequencies during the contest. Usually, I'll make a number of contacts there and it is a chance to generate some interest in contesting with the folks that hang out on FM. It is also common to run into operators that have 2-meter SSB but can only run FM on 70 cm, so we'll switch over to 446.0 MHz to make an additional contact. On the other hand, I have only FM gear on 222 MHz, so I ask SSB operators to go to an FM simplex frequency and work me via FM. Sometimes, I bring along a horizontally-polarized Yagi (not common on FM) for making these contacts so that we don't suffer the signal loss associated with cross-polarization. For weak-signal work, FM is FAR inferior to SSB and CW, but it may be a way for you to add an additional band to your station.

Summary

Rover operation is FUN. It starts out with the fun challenge of equipping your station and figuring out how to make it work. The actual event is fun because it gets you out of the shack, out of the house and cruising down the highway. Instead of being just another station in Denver or Colorado Springs, now you are a new grid. Once you give this a try, you'll find all kinds of ideas on how to approach your next rover operation. So give rover operation a try and have some fun.

I think that it is only fair to warn you that roving can be addictive and your family and friends may try to obtain professional help for you.

References

RoverLog website – http://roverlog.2ub.org/

9. VHF QRP

Mention "QRP operation" and most radio amateurs think of a small CW transceiver for the HF bands. Mention "VHF QRP" and the response may be more like "what's the matter, your transmitter broke?" The surge of interest in QRP is largely focused on the HF bands and most weak-signal operation on the VHF bands is high power, for good reason. Putting together a competitive weak-signal station requires careful attention to every decibel in the system — receive sensitivity, transmission line loss, antenna gain and, yes, transmitter power. On the other hand, there is something to be learned from the QRP community about having fun with amateur radio.

What is QRP?

QRP is normally defined as operating with 5 W of output power or less. If you dig a little deeper, you'll find that low-power operation carries a lot more with it.

QRP is generally associated with:

- Compact, portable, battery-powered equipment (often used portable in the outdoors)
- A personal challenge and/or a minimalist approach (get the job done using efficient equipment)
- Emphasis on operator skill (especially CW operation)
- Contesting or other events that promote QRP activity

Are these elements of radio operating relevant to VHF and up?

Our VHF equipment is not always compact and portable but in recent years there has been a significant reduction in the size and weight of VHF and up equipment. It started with the combination HF/VHF/UHF mobile rigs from ICOM (IC-706) and Yaesu (FT-100). For QRP enthusiasts, the FT-817 from Yaesu is a backpack-ready 5 W rig that spans HF through 70 cm. (Unfortunately, none of these rigs include the 222 MHz band.) Now that I think of it, we should include the older single-band all-mode rigs such as the FT-290R and the IC-502. We see hams using these rigs for VHF/UHF mountaintop and grid expeditions. So, yes, there is a match between compact, portable operation and VHF and up.

With regard to taking on a personal challenge in radio operating, the weak-signal VHF/UHF enthusiasts are already there. The higher bands were once thought to be of no use except for limited line-of-site propagation. The weak-signal ham community has proved that idea wrong.

The weak-signal VHF world also puts an emphasis on operator skill, including the use of CW. You must be a fast and efficient radio operator to make contacts when VHF band conditions are marginal or changing rapidly. Most serious VHF operators have had the experience of trying to work a distant station on SSB, then switching over to CW to complete the contact. Whether you like using CW or not, it does get through tough conditions better than phone, so it is important to have it in your bag of tricks.

With regard to contests, this is where QRP VHF is formally established. The major VHF contests have a special entry category for QRP operation, with a maximum power level of 10 W, not 5 W. The ARRL contests refer to this category as "Single Operator Portable" while the CQ World-Wide VHF contest just calls it "QRP." The intent of these categories is to encourage portable operation, presumably from a rare grid or mountaintop location.

In summary, we can check the box on all of the main QRP elements as applying to VHF and higher.

Intro to VHF QRP

First, let's talk about some of the more popular frequency bands above 50 MHz. I am going to discuss the bands up through 70 cm because they are the most commonly used. Certainly, there is a huge amount of spectrum above 70 cm with lots of potential, especially if you are into experimentation and homebrewing of equipment.

Band	Frequency Range	Calling Frequencies	Comments
6m	50 to 54 MHz	SSB: 50.125 MHz FM: 52.525 MHz	Normally local communications but sporadic-E and F2 can produce skywave propagation.
2m	144 to 148 MHz	SSB: 144.200 MHz FM: 146.52 MHz	The most popular VHF band, used for local communication via simplex and repeaters. Normally local communications but long distance tropo propagation and sporadic-E propagation occasionally occurs.
1.25m	222 to 225 MHz	SSB: 222.100 MHz FM: 223.5 MHz	Propagation similar to 2m but rarely sporadic-E
70cm	420 MHz to 450 MHz	SSB: 432.100 MHz FM: 446.0 MHz	Propagation similar to 2m but no sporadic-E

Note: This table shows the US amateur bands, other countries may have different frequency allocations.

Equipment

There are a number of transceivers available for VHF QRP. One of the most exciting rigs to appear on the scene is the FT-817 (and the newer model, FT-818), which covers HF, 6m, 2m and 70 cm. The Elecraft KX-3 QRP rig includes the 6-meter band standard and has an option for the 2-meter band. There are also the older single-band radios available on the used market, such as the FT-290R series from Yaesu and the IC-202 series from ICOM.

Antenna

The antenna for any amateur radio station is a critical component. The key difference at VHF and higher is the shorter wavelength, which means antenna elements are much shorter and perhaps more numerous. The polarization of the antenna is important under most situations, since you want to have your antenna with the same polarization (horizontal or vertical) as the station you are contacting. Most FM-oriented stations use vertical polarization, consistent with easy mobile mounting and simple omni-directional antennas. Serious weak-signal VHFers almost always use SSB or CW and horizontal antennas.

For the 6-meter band, we can adapt many of the standard HF wire antenna designs. For example, the classic half-wave dipole is a good choice, providing an efficient radiator about 9 ½ feet long. Yagi antennas for 6 meters are normally mounted horizontal and provide significant gain over a dipole.

For the 2-meter band, the wavelength and antenna elements are quite short (about 1 meter or 39 inches) compared to the typical HF antenna. Here, the Yagi antenna is the most popular choice, with as few as 3 elements and as many as 17 elements. For CW or SSB work on this band, you'll definitely want to be horizontally polarized.

The bands higher than 2 meters tend to also use Yagi antennas but with correspondingly shorter elements. The shorter wavelength allows for more antenna gain within the same boom length. Of course, a directional antenna means that you need a method for pointing the antenna in the desired direction.

Transmission Lines

Transmission line loss is an issue at VHF and higher frequencies. This loss is usually specified in dB per 100 feet of cable length. The losses of some common coaxial transmission lines are shown in the table below. Small cables such as RG-58 have high loss at VHF frequencies, losing 3 dB (half the power) in 100 feet at 50 MHz. However, RG-58 might be acceptable for shorter cable runs, say 25 feet or less. RG-8x is not much larger in diameter but delivers lower loss. For longer runs, the larger "full size" RG-8U and 9913 are necessary to control transmission line loss.

Cable Type	Loss per 100 feet (50 MHz)	Loss per 100 feet (150 MHz)	Loss per 100 feet (450 MHz)
RG-58/U	3.1 dB	6.2 dB	10.6 dB
RG-8x	2.3 dB	4.7 dB	8.6 dB
RG-8U/foam	1.2 dB	2.3 dB	4.7 dB
9913FX	0.9 dB	1.6 dB	2.8 dB

Source: Cable X-perts, Inc. catalog, http://www.cablexperts.com

Operating

The place to start on VHF is to go to the calling frequency and call CQ. Unlike the HF bands, there is a tendency to mix CW and SSB operation on the same frequency space on VHF. For example, on 2 meters you might hear an SSB signal calling CQ on 144.200 MHz then a few minutes later hear a CQ using CW. Someone might even respond to the CW CQ using SSB and take up the QSO on phone. It is good operating practice to move off of the calling frequency once contact is established. However, you'll hear people rag-chewing on the calling frequency, especially in areas that have little VHF activity.

Without a band opening, you are dependent on local activity to make contacts on the VHF bands. Local activity is, well, local and depends on how many VHF operators there are in your area and how often they get on the air. The amount of activity on these bands will vary dramatically from place to place. Some areas have a formal or informal activity night, sometimes by band. For example, Monday night may be the 2-meter activity night, while Tuesdays might be for 1.25 meters. Obviously, this is a good time to get on the air, check out your equipment and work some of the local VHF crowd. Another opportunity is a local VHF SSB net, most common on 2 meters.

Finally, VHF contests are great for concentrating activity and represent a chance to work lots of stations in a short period of time. These weekends are my favorite weekends to operate VHF, not so much to compete in the contest but to enjoy the higher level of activity on the bands.

Mountaintop Operating

One of the advantages of QRP operating is that the equipment is small and portable. You can operate from almost anywhere! For VHF, the obvious thing to do is increase your *height above average terrain*. In other words, drive or hike to your favorite mountaintop location and activate it. In Colorado, we have an event designed just for this purpose, called the Colorado 14er Event. VHF contests are an excellent opportunity to operate from your local high spot. Summits On The Air (SOTA) is a fun program for mountaintop operating. If you don't have a mountain close at hand, then check out the local fire lookout tower, lighthouse, grain silo or skyscraper.

References

Colorado 14er Event – http://www.ham14er.org/

Summits On The Air (SOTA) – https://www.sota.org.uk/

PART II
BLOG POSTS

This section contains selected postings from the k0nr.com weblog. They have been edited and updated for inclusion in this book.

10. Choose Your 2m Frequency Wisely

You've just purchased your first 2-meter FM transceiver and have been chatting with both old and new friends around town on the 2-meter band. You and your buddies decide to find an out of the way frequency to hang out on. After tuning around, you find a nice, quiet frequency that no one seems to be using and start operating there. Nothing to worry about, right?

Not so fast, there are a few more things to consider when selecting a frequency on the 2-meter band. Let's take a look at the key issues.

FCC Rules

The first thing we need to know are the frequencies that the FCC has authorized for our particular license class. For the HF bands, the frequency privileges depend greatly on the license class of the operator. Above 50 MHz, the frequency allocations are the same for Technician licenses and higher. In particular, the 2-meter band extends from 144 MHz to 148 MHz. The FCC Rules say that any mode (FM, AM, SSB, CW, etc.) can be used on the band from 144.100 to 148.000 MHz. The FCC has restricted 144.0 to 144.100 MHz to CW operation only.

Band Plans

Knowing the FCC frequency authorizations is a good start but we need to check a bit further. Amateur radio operators use a variety of modulation techniques to carry out communications. Often, these modulation techniques and operating techniques are incompatible since a signal of one type can't be received by a radio set to another modulation type. For example, an SSB signal can't be received on an FM receiver (and vice versa). We need to use our authorized frequencies wisely by sharing the band with other users and avoiding unnecessary interference. Thus, it makes sense to have a *band plan* that divides the band up into segments for each type of operation.

Two-meter Band Plan

As shown in the table, the 2-meter band plan supports a wide variety of radio operation. Large portions of the band are dedicated to FM operation, consistent with the popularity of the FM mode. There are portions of the band designated for repeater *outputs* (which is the frequency that we tune to receive the repeater) and repeater *inputs* (which is the frequency we transmit on to use the repeater). Notice that these segments are positioned 600 kHz apart consistent with the standard 2-meter repeater offset. There are also frequencies designated for FM simplex.

On the low end of the band, we see segments for some of the more exotic modes. At the very bottom is the CW portion, which includes Earth-Moon-Earth (EME) operation. EME operators communicate by bouncing their signals off the moon. Further up the band, we see segments for SSB operation and beacon operation. SSB is the preferred voice mode for so-called "weak signal" operators. Beacons are transmitters that are always on, transmitting a short CW message as a propagation indicator for distant stations. We often think of 2 meters as a local coverage band but when conditions are right, contacts can be made with stations over a thousand miles away. Of course, conditions are not always right so having a beacon on the other end of the desired communication path lets you know how propagation is in that direction.

2m Band Plan
As approved by the ARRL VHF-UHF Advisory Committee, simplified by KØNR to reflect usage in Colorado.

The Colorado Council of Amateur Radio Clubs (CCARC) publishes the official *2 Meter Frequency Use Plan* for the 2 Meter band in Colorado.

Frequency	Usage
144.000-144.100	CW
144.100-144.275	Single-sideband (SSB Calling Frequency = 144.200)
144.275-144.300	Propagation Beacons
144.300-144.500	OSCAR (satellite) APRS Frequency = 144.390 MHz
144.500-144.900	FM Repeater Inputs
144.900-145.100	Packet Radio
145.100-145.500	FM Repeater Outputs
145.500-145.800	Misc. and experimental modes
145.800-146.000	OSCAR (satellite)
146.010-146.385	FM Repeater Inputs
146.400-146.595	FM Simplex (National Simplex Calling Frequency = 146.52 MHz)
146.610-147.390	FM Repeater Outputs
147.405-147.585	FM Simplex
147.600-147.990	FM Repeater Inputs

Note: The 2m FM channel spacing in Colorado is 15 kHz (repeaters and simplex).

Radio amateurs also use 2 meters for OSCAR satellite operation, sending signals *to* a satellite (uplink) or receiving signals *from* the satellite (downlink). The OSCAR segments don't specify a particular modulation type since CW, SSB and FM are all used for OSCAR operation. Because of their elevation above the earth, satellites can hear signals from all over the US simultaneously, so they are very susceptible to interference.

Most of this non-FM operation can be easily interfered with by signals from other users. EME signals, for example, are usually quite small since the signal has to make the round trip from the earth to the moon and back. If a local FM operator fires up in the EME portion of the band, an EME signal that can't be heard by an FM receiver can be wiped out by the FM signal. Similarly, an operator chatting across town on 2 meters could interfere with a satellite hundreds of miles away and not know it. This is particularly a problem with FM receivers, which won't even notice low level CW and SSB signals.

FM Operating

The most common 2-meter rigs are basic FM mobile or handheld transceivers. These radios usually tune the entire 2-meter band from 144 MHz to 148 MHz in 5-kHz steps. The band plan indicates the proper range of frequencies for FM operation but there is more to the story. FM operation is "channelized", meaning that specific 2-meter FM frequencies are identified by the band plan. The use of channels is especially important for repeaters because they don't easily move around in frequency and are coordinated to minimize interference. The idea is to have all stations use frequencies that are spaced just far enough apart to accommodate the signal without interfering with the adjacent channels.

You might think that the spacing between channels would be 5 kHz, which is the tuning step of most FM radios. This doesn't work because an FM signal occupies a bandwidth that is more than 5 kHz wide. Even though we talk about a signal being on a specific frequency, the signal actually spills out on either side of the frequency by about 8 kHz. This means that a typical FM signal is about 16 kHz wide.

(You may recall that amateur 2m FM uses ±5 kHz frequency deviation. So doesn't this mean the bandwidth is 10 kHz? No, it doesn't quite work that way and the signal is actually wider than 10 kHz. I might be able to show the math behind this but it makes my head hurt. Perhaps in some future article.)

The channel spacing needs to be at least as wide as the bandwidth of the signal, which allows room for each signal without interfering with the adjacent channel. In Colorado, the channel spacing is 15 kHz, which is a bit tight for our 16-kHz-wide signal. In other parts of the country, a 20-kHz spacing has been adopted to provide for more separation between channels. Obviously, you get more channels on the band with 15-kHz spacing than with 20 kHz, but you have to put up with more adjacent channel problems.

When using a repeater, you just need to dial in the published repeater frequency and set the transmit offset, either + 600 kHz or − 600 kHz. Most modern 2-meter radios automatically take care of setting the proper offset (based on the band plan). If you need to set the offset manually, the rule is very simple. If a repeater's output frequency is in the 147 MHz range, it uses a + 600 kHz offset. Otherwise, it requires a − 600 kHz offset. For repeaters that require a CTCSS tone for repeater access, you will have to set the proper tone frequency on transmit.

> **The *National Simplex Calling Frequency* is 146.52 MHz. (Also referred to as simply the *Calling Frequency*.)**

For simplex operation, the standard simplex frequencies listed in the table below should be used. These simplex frequencies are grouped in the 146 MHz and 147 MHz range as listed in the table below.

2m FM Simplex Frequencies
Colorado Band Plan

146 MHz Range	146.400, 146.415, 146.430, 146.445, 146.460, 146.475, 146.490, 146.505, **146.520**, 146.535, 146.550, 146.565, 146.580, 146.595
147 MHz Range	147.405, 147.420, 147.435, 147.450, 147.465, 147.480, 147.495, 147.510, 147.525, 147.540, 147.555, 147.570, 147.585

In regions that use a 20-kHz channel spacing, you'll find simplex frequencies in the same general frequency range but with different spacing. For example, standard 2-meter simplex frequencies might be 146.520, 146.540, 146.560, etc.

The FCC View on Band Plans

Sometimes I hear radio amateurs say, "Band plans are voluntary so I don't need to pay any attention to them. I can do whatever I want as long as I don't break the FCC rules." Unfortunately, such an attitude does not promote efficient use and sharing of the amateur bands. Imagine the chaos on the ham bands if everyone took this approach. It also may be a violation of FCC rules.

On Oct 18, 2000, in a ruling concerning a repeater operator's failure to conform to the prevailing band plan, FCC Special Counsel for Amateur Radio Enforcement, Riley Hollingsworth commented on the issue. He said "Band plans minimize the necessity for Commission intervention in Amateur operations and the use of Commission resources to resolve amateur interference problems. When such plans are not followed and harmful interference results, we expect very substantial justification to be provided, and we expect that justification to be consistent with Section 97.101."

Section 97.101 is the part of the FCC rules that says (among other things):

- In all respects not specifically covered by FCC Rules each amateur station must be operated in accordance with good engineering and good amateur practice.
- Each station licensee and each control operator must cooperate in selecting transmitting channels and in making the most effective use of the amateur service frequencies.

The FCC has clearly stated that they expect hams to share the bands by following accepted band plans. More importantly, this is the right thing to do for the benefit of the amateur radio service.

Summary

The fine points of the band plan can be a bit confusing. However, a few simple guidelines can help, especially if you are operating only FM.

- FM voice simplex and repeater operation should occur only above 145.100 MHz (and only in the OSCAR subband if you are working an FM satellite)
- When operating through a repeater, make sure you are tuned to the published repeater frequency with the proper transmit offset.
- When operating simplex, use a simplex frequency designated by the local band plan.

We've only covered the 2-meter band in this article. If you are operating on other bands, be sure to check the appropriate band plan before transmitting. Note that this article is written for amateur radio operation in Colorado and that other locations may have a different band plan.

References

What Frequency Do I Use on 2 Meters?
http://www.hamradioschool.com/what-frequency-do-i-use-on-2-meters/

What Frequency Do I Use on 70 Centimeters?
http://www.hamradioschool.com/what-frequency-do-i-use-on-70-centimeters/

11. FOT: Frequency, Offset and Tone

One question I often hear from new hams (and maybe some not-so-new hams) is "why can't I get into the repeater?" They get their hands on a new radio, set it up to use one of the local repeaters and it's not working. Now what?

There can be a whole bunch of reasons why you can't get into a repeater so it is difficult to come up with a quick fix for all situations. However, in this article we'll talk about some basic troubleshooting steps to help diagnose the problem. For this article, I am assuming that your first rig is a handheld VHF/UHF transceiver but the general approach will work with mobile or base transceivers, too.

FOT

Many times the problem is due to not having the transceiver programmed correctly. The key things we have to pay attention to are: Frequency, Offset and Tone (**FOT**). To access a repeater you need to have its **Frequency** entered into your radio, have its transmit **Offset** set correctly and have the right CTCSS **Tone** turned on. You might not need to check all of these things in that exact order but it is a good way to approach the problem. Using the programming software (and suitable cable) for your radio can be a big help.

Frequency –First you need to program in the frequency of the repeater you want to access. The actual key strokes or knob turns will depend on the particular model of radio so consult your operating manual. The frequency you enter is the *repeater transmit frequency* which will be your *receive frequency*. Repeaters are always referred to by their transmit frequency, which can be found in an online or printed repeater directory.

Offset – Next, we need to make sure the proper transmit offset is programmed into the radio. This is the difference in frequency between the repeater transmit frequency and its receive frequency. Your transceiver will automatically shift your frequency when you transmit, *if* you have the right offset programmed. In most parts of the US, the standard offset is 600 kHz on the 2m band and 5 MHz on the 70cm band, and can be either in the positive (+) or negative (-) direction. Your repeater directory will list the offset and direction. Most radios will default to the standard offset but you may have to select + or – offset. Usually a + or – symbol will appear in the display to indicate the offset selected.

As an example, my repeater is on 447.725 MHz with a -5 MHz offset. So you would enter 447.725 MHz into your radio, make sure the offset is set to 5 MHz and select "–" as the offset direction. You can verify that your radio is programmed correctly if you see 447.725 MHz displayed during receive, which should change to 442.725 MHz when you push the transmit button.

Some radios do not use the concept of transmit offset but instead require you to enter the transmit frequency directly. You will need to calculate the transmit frequency (using a 600 kHz or 5 MHz offset) and program it into your radio.

Tone – For most repeaters, you will need to transmit a CTCSS tone to access the repeater. (CTCSS is Continuous Tone Coded Squelch System.) Repeaters with *carrier access* do not require a tone, so you can skip this step. This is normally a two-step process: set the tone frequency and then enable the tone. Sometimes this is done with one selection (with "Off" being an option for the tone frequency). Some radios have separate settings for the transmit tone and receive tone. For now, just leave the receive tone off, since it can be a source of confusion. The tone that you need to set is your *transmit tone*. Most radios display a "T" somewhere on the display when the tone is enabled. Again, check your operating manual.

Kerchunk

At this point, you should be ready to try accessing the repeater. After listening on frequency for a minute, transmit and identify using your call sign. On most repeaters, you will hear a short transmission coming back from the repeater along with a courtesy beep. A courtesy beep is just a short audio tone or tone sequence that occurs after someone finishes transmitting. If you hear the beep, then you accessed the repeater. Congratulations! Go ahead and make a call and see if someone will come back to you.

Troubleshooting

What if you don't hear the repeater coming back to you? Then we need to go into troubleshooting mode. If the radio is new, you might wonder if it is even working properly. The quality level of today's equipment is quite good, so most likely your radio is just fine. Still, you may need to check it out.

First, you can check to make sure your radio is receiving properly. In the US, a good way to do this is to tune into your local NOAA weather transmitter. These transmitters are on the air continuously, operating on 162.400, 162.425, 162.450, 162.475, 162.500, 162.525 or 162.550 MHz. These frequencies are outside of the 2-meter ham band but most ham transceivers are able to listen to these frequencies. You'll want to set this frequency as simply as possible...use the keypad or VFO mode to enter it directly. In most cases, you can just try the short list of frequencies until you hear the transmitter in your area.

Next, you might want to know that your radio is able to transmit a signal. The best way to do this is find a local ham nearby that can run a simplex check with you. By nearby, I mean within 5 miles or so, because we want someone so close that there is no question about whether they should be able to contact us. Program your radio to a 2-meter simplex frequency such as 146.52 MHz (the 2m FM Calling Frequency). For this test, we do NOT want the transmit offset turned on…the radio needs to be set to *simplex*. You can double check this by looking at the display when transmitting—it should show 146.52 MHz (transmit frequency is the same as the receive frequency). For this test, we don't care about the transmit tone…it can be on or off. Have the other ham give you a call and see if you can contact him. If you happen to have a second transceiver, you can try this test yourself – just see if each radio can hear the other one. One warning: do this on a simplex frequency. Trying to go through a repeater can really confuse things because you may not have the offset and tone set properly. Even more confusing is that one radio can "desense" the other radio, which means that the other radio's receiver will be overloaded and not able to receive the repeater's signal. Using simplex keeps things simple.

The final thing to check is whether your signal is able to reach the repeater. Well, that is a bit of a challenge! For starters, are you sure you are within range of the repeater? Have you ever heard a signal from this repeater, and was it full scale on your S meter? You may want to ask local hams about whether you should be able to hit the repeater from your location *with the radio you are using*. For that matter, you might want to check if the repeater is actually on the air – they do go down from time to time.

This brings us to an important point about the use of handheld transceivers. They are really, really handy. How else can you carry a complete ham radio station in your hand? Well, the tradeoff is that an HT operates with relatively low power (5 W or less) and has a compromised antenna. (The standard *rubber duck antenna* on an HT is a *very convenient crummy antenna*.) You may need to add some extra umph to your signal by improving the antenna. Some good dual-band choices are a longer whip such as the Diamond RH77CA or a magnetic-mount mobile antenna placed on a vehicle or other metal object.

Summary

In this article, I've tried to provide some assistance in figuring out why you aren't hitting the repeater. The most common problem for newly acquired radios is getting them programmed correctly (**remember FOT: Frequency Offset and Tone**). Once you have that right, it is usually just making sure that you have enough signal to make it to the repeater.

73, Bob K0NR

Note: this article is adapted from Hey, Why Can't I Access the Repeater? on hamradioschool.com

12. VHF Grid Locators

Amateur radio operators on the VHF bands use the Maidenhead Locator System for indicating their location, often referred to as a *grid locator*. A grid is defined by 1° latitude by 2° longitude, measuring approximately 70 × 100 miles in the continental US. A grid is indicated by two letters (the field) and two numbers (the square). For example, the home location of KØNR is located in grid DM79, which includes greater Denver. Greater Colorado Springs is to the south in grid DM78. These things are commonly called *grid squares* even though they are really *rectangles* or perhaps *trapezoids*. (Actually, since the earth's surface is curved, they are more complicated three dimensional shapes.) Anyway, I try not to say *grid square*, but it often sneaks in.

World map showing the first two characters of the grid locator. (Graphic: EI8IC website)

The most commonly used version of grid locator is the 4-character version (e.g., DM79), which is used for the major VHF contests and awards. Be aware that there is also a 6-character version, which provides a more precise location. The 6-character locator is used in some contests including the ARRL 10-GHz and Up Contest. The 6-character locator adds on two additional characters to the 4-character version, so they are consistent and compatible. For example, my 4-character locator is DM79 and my 6-character locator is DM79nc.

Finding Your Grid

If you know your latitude and longitude, you can use AMSAT's conversion page to find your grid locator. Arguably the more useful set of tools are on the K2DSL page, which allows you to enter your physical address to pull up a map of your grid locator. You can also just enter a call sign and have the address pulled automatically from the QRZ database.

EI8IC has a grid map of the entire world and other useful mapping tools on his website.

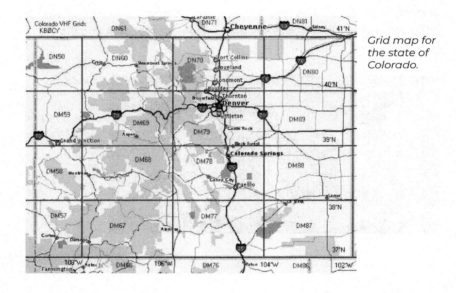

Grid map for the state of Colorado.

For locations in Colorado, the map shown above can be used to determine the grid. This map is a rough approximation of where the grid lines are located. If you are near a grid boundary, you need to determine your grid via more accurate methods. In particular, note that the eastern boundary of grids DN80, DM89, DM88 and DM87 is 2 to 3 miles east of the Colorado state line.

References

Maidenhead Locator (Wikipedia) – https://en.wikipedia.org/wiki/Maidenhead_Locator_System

AMSAT Grid Locator Calculator – http://www.amsat.org/amsat/toys/gridconv.html

K2DSL Grid Locator Calculator – http://www.levinecentral.com/ham/grid_square.php

EI8IC Ham Radio Resources – http://www.mapability.com/ei8ic/

13. Rescue on Uncompahgre Peak

I came across this story in my archives, written by me way back in August 1992. This was before mobile phones were commonly available, so ham radio turned out to be critical in this incident. Even today, there are many places in the Colorado backcountry where mobile phones don't work but amateur radio can communicate. My call sign at the time was KBØCY

Something happened on the way to Uncompahgre Peak on August 8, 1992.

Around noon, my brother, my two nephews and I made it to the summit and had just signed the log. I called on 146.52 and contacted Chris, NQ5V, who was somewhere to the east of me (Creede, I think). This must be his summer location, since his callbook address is Texas. We talked about the trail up Uncompahgre, since he was interested in hiking it.

Find the Radio Guy

After I signed clear with NQ5V and was about to start down the mountain, a teenage boy came up to me and said he had been sent to "find the guy with the radio" because a girl had been hit by a rock down below and was hurt. I am not sure how they knew I had a radio, other than I used it once on the way up the trail. The story seemed rather sketchy and I was skeptical but asked NQ5V to standby on frequency because we may have a medical emergency. At that time, Arnold/W7JRC, from Cedaredge, CO, came on frequency and said he had a phone nearby. (NQ5V did not have a phone available.) A second, older teenager came up the the trail with more information. He said he was a pre-med student and had search and rescue experience. He had more detailed info which made the story more clear. At this time, I concluded that we had a real emergency and asked W7JRC to call the authorities. I handed my HT to the older teenager and had him describe the victim's condition to W7JRC. W7JRC had some trouble contacting the police, but eventually got through to the Ouray County Sheriff's Office. (It turned out we were in Hinsdale County, but we did not know that at the time.)

Jim/NR5Y (also close to Creede, I think) came on frequency and said that he was close to a telephone. I was not always able to communicate with NR5Y, so NQ5V relayed to NR5Y. Since W7JRC was having trouble with getting the telephone call through, I asked NR5Y to also try to place a call. He called the Mineral County Sheriff, who relayed to Hinsdale County. All this time, I was moving down the mountain to try to get closer to the victim without losing my radio contact. About this time, my HT battery went dead, so I switched to my spare. (Good thing I had one!) As I moved onto the saddle below Uncompaghre, I lost contact with W7JRC and contact with NQ5V got much worse, but usable. About this time, Doug/NØLAY, came on the air and his signal was very strong at my location which allowed me to stay on low power and conserve my HT batteries. NØLAY apparently came on the air in response to a call from the Hinsdale County Sheriff. NØLAY also had a radio which was on the sheriff's frequency and relayed information from me to the sheriff's dispatch.

I had not proceeded down any further because I was certain that I would lose radio contact with NØLAY. The victim had several people with her that had First Aid training and was about 1000 feet below me at the bottom of a cirque. I sent the older teenager back down to the victim with instructions to signal me as to her condition. We both had signal whistles – two whistles meant her condition was the same (stable), three whistles meant her condition had deteriorated. After I got the two whistles back, I felt like things were going to be OK.

About that time, NØLAY relayed that an ambulance had been dispatched to the trailhead and a search and rescue (S&R) person was on the way up the trail with a trail bike. Also, a helicopter had been dispatched from Montrose. It took us a little while to communicate to the sheriff where the victim was, but we had a pretty good topo map, so we eventually gave them an accurate fix on the location. As I was listening to NØLAY relay, I realized that my Kenwood TH-77A could receive most police frequencies. NØLAY provided me with the frequency and I programmed it into the HT, scanning between 146.52 and the sheriff's frequency. This allowed us to listen in on what was going on. In fact, many times I was clearly hearing the various parties while they were having trouble communicating.

Search and Rescue

The S&R guy on the trail bike made it to the accident scene without us noticing him. He had parked his bike about half a mile away from us and had scrambled down to the victim. The first time I was aware of his position was when he transmitted from the accident site. He confirmed that the girl was pretty bashed up, but stable, and needed a helicopter ride out. About this time, the sheriff's dispatch reported that the helicopter was about 5 minutes out (I think it turned out to be more like 15 minutes away). Soon the helicopter came up on the sheriff's frequency and I could hear the S&R guy coordinating with the helicopter pilot. The two-seater helicopter landed and they put the girl in the second seat. Apparently, she was stable enough to walk to the helicopter with some assistance. The alternative was to put her outside the chopper in a litter. The helicopter lifted off and set back down a few minutes later near the ambulance which was near the trailhead. The two-seater chopper was not a medical evacuation helicopter and the plan was that Flight-For-Life from Grand Junction would pick up the victim at the ambulance location. It turned out that Flight-For-Life was unavailable so they took the victim to a hospital by ambulance (to a local clinic, then Gunnison, I think).

We stayed on the ridge until the chopper headed for home, then we did the same. On the way down, the S&R guy on the trail bike caught up with us and we talked about the accident. He said the girl lost some teeth, had facial cuts, internal bleeding and swelling in the face, but was in stable condition. He said that without the radio report that they would be just getting the initial call at the time he was heading home. That is, we saved about 5 hours on the response time with amateur radio.

I have carried my HT on every 14er hike I have ever done and had considered the possibility of using it for emergency communications. I guess I never gave it too much thought because people venturing into the backcountry need to have a self-sufficient attitude. That means being prepared and preventing or handling any emergency situation on your own. But the unexpected happens, and here I was in the middle of a medical emergency. It certainly has caused me to take this emergency communications thing more seriously.

Lessons Learned

Things I learned that day:

- Always carry an extra HT battery (or two)
- Always carry a decent portable antenna (more than a rubber duck)
- Always carry a good topo map, even if you don't need it to follow the trail.
- Make note of what county you are hiking in when in unfamiliar parts of the state. This aids in getting to the right Sheriff's office. (This is important because the person you contact via radio is likely to be two or three counties away.)
- My signal whistle (which has caused considerable abuse from a

few hiking companions) is actually useful.
- Extended coverage receive is very useful in emergencies. (I am still thinking about extended transmit — I clearly could have used it in this case.)

I was very pleased that everyone reacted quickly but in a professional manner. The radio amateurs all helped out when they could but stayed out of the way when appropriate. I am sure we can find some things that could have been done better, but I felt like things went well overall.

14. The Use of 146.52 MHz

One of the local radio clubs recently had a heated discussion about the use of the *National Simplex Calling Frequency,* which is 146.52 MHz in the US (per the ARRL band plan). You have probably heard the argument before.....is the calling frequency reserved to just calling or is it OK to ragchew on that frequency?

In the Summer 2006 issue of *CQ VHF*, I wrote about mountaintop operating and included these thoughts on the use of 146.52 MHz:

Calling Frequency

What frequency are you going to use to call CQ from your favorite high spot? Well, the calling frequency, of course.... most likely 146.52 MHz. This usually works pretty well as many simplex-oriented operators make it a point to listen on .52. While we don't normally make long CQ calls on VHF FM, making a call such as "CQ Five Two, this is K Ø N R on Pikes Peak" is a good way to go.

One problem I've run into is when the calling frequency is tied up with lengthy contacts by other hams. If the frequency is in use, I generally just stand by and wait for them to finish. If it seems appropriate, I might break in and chat with them.

> *Disclaimer: It is difficult to write authoritatively in a national ham magazine about VHF issues that often tend to be regional in nature. What works in rural areas with lower pop-*

ulation density may not apply in New York City. Ignoring that, I'll jump in with both feet (maybe with one in my mouth, who knows?)

What is the purpose of a calling frequency? Back in the old days of crystal-controlled rigs it was important that we had common channels crystalled so we could talk to each other. We typically only had a dozen or so channels, so having a common calling or simplex frequency (or two) was an obvious thing to do. These days, we have synthesized 2m FM rigs that cover the entire 4 MHz band in 5 kHz steps. Now the purpose of a calling frequency is, well, for calling. You use 146.52 MHz when you want to establish contact on the band, lacking any other information. For example, if I know my buddy Steve/KØSRW is going to be listening on the 146.94 MHz repeater, I'll call him there. If I know the local DX crew hangs on out 146.46 MHz, I'll make a call there. But when I don't have any other information, and I am making a call or listening for a call, I go to the calling frequency. Why? Because that's what it's for! If I am out of repeater range and I just want to talk to someone on simplex, I try the calling frequency.

The Three Minute Rule

There are two ways to make a calling frequency useless:

> 1. No one ever uses the calling frequency (nobody there, nobody home)
> 2. The calling frequency is always tied up due to lengthy contacts

So we need to encourage hams to monitor and use the calling frequency, but not monopolize it. We don't have to be extreme about it. Perhaps we can adopt a "three-minute rule of thumb": if I am in a contact with another station on the calling frequency for more than 3 minutes, it is time to change to a different frequency. This opens up the frequency for other hams to use. Just as important it keeps the long ragchew sessions away from the calling frequency. These long sessions have a tendency to discourage monitoring of 146.52 MHz.

One ham recently told me that he tries to keep a receiver tuned to .52 for anyone just passing through the area that might need some help. But when some of the locals get on the frequency and chat for an hour, the radio gets turned off.

There, I said it: *the calling frequency is for calling, not for ragchewing.*

> **Monitor and use the 2-meter calling frequency, but don't monopolize it with extended radio contacts.**

15. Paul Rinaldo's Rule

Recently on the AMSAT-BB email list, there was a discussion about some new satellites about to be launched. Some folks criticized the implementation of the satellite hardware (as in "they should have done *this* instead of *that*.") Bob Bruninga, WB4APR, posted an excellent response that bears repeating:

> **Paul Rinaldo's rule of Amateur Radio Progress:**
>
> *Progress is made in Amateur Radio by letting energetic individuals move forward. Conversely, nothing in Amateur Radio is accomplished by complaining about other individual's projects. Simple summary: If you don't like their project, then go do or support your own choices. Get out of their way.*
>
> *The service is where we are allowed to experiment as individuals. This means if you have an experiment, then do it. If someone else has an experiment you like, then contribute to it, support it, or get out of the way. It's the individuals that move forward and do something that makes Ham radio progress.*
>
> *Conversely, It's all the naysayers, thought police, kibitzers, complainers, arm chair lawyers and couch potatoes that hold much progress back. Not one cloud of their hot air will move anything forward. The only thing it does is make us all look like old fuds and many of the would-be progressives just throw in the towel and instead of a great hobby, they go get a real job instead.*
>
> *So, again, Ham radio is an unbelievably diverse collection of intersts, modes, techniques, applications, projects, missions, activities, directions, places, groups, frequencies, bands, devices, propagation, tests, events and experiments. Choose those you are interested in, jump in, contribute, move forward... Do not waste your time (and other's) trying to hold others back from their interests.*

NOTHING IS EVER ACCOMPLISHED PROGRESSIVELY BY TRYING TO FORCE OTHERS INTO SPENDING THEIR TIME ON YOUR VIEW OF AMATEUR RADIO.

The best you can do is find other people that actually do something and support them in a direction you want and hang on for the ride. I call this **Paul Rinaldo's Rule of Amateur Radio**, because I learned it back in the late 1970s or so when I was on the board of directors and Paul was president of AMRAD which was working on the AX.25 spec, and trying to develop the early East Coast Packet System. We all met frequently, and everyone had ideas of which way to go. It's the sentence that begins with "what they need to do is..." that Paul pointed out was pointless.

Or something like that. Over the years, in all the clubs and organizations of Amateur Radio, I find it the rule to live by. If you have an idea, do it. If someone else has an idea, either join it or support it. If it's a dumb idea, it will die, don't waste your time trying to assure its demise. But complaining about others people's progress just makes no sense to me.

Bob, WB4APR

16. Don't Get Stuck On 2 Meters

When I first got started in amateur radio (many years ago), one of the engineers that I was working with at a summer job told me "Don't get stuck on 2 meter FM." At the time I was a college student and felt lucky enough to have 1) found time to pass my Novice exam, including Morse Code test, 2) found time to travel 150 miles to the regional FCC office and pass my Technician exam, and 3) scraped up enough money to buy a basic 2m FM mobile rig. I was in Technician ham heaven, playing around on 2m FM, both simplex and repeaters. Oh, and we had this cool thing called *autopatch* that let you make actual phone calls from your car. I really wasn't worried about getting "stuck on two."

Even though my discussions with this ~~Old Fart~~ Experienced Radio Amateur revealed that he didn't see 2-meter FM as *Real Ham Radio*, I could see that he had a point. Two-meter FM is only a small part of the ham radio universe and it would be easy to just hang out there and miss out on a lot of other things. I was reminded of this recently by K3NG's post: Things I Wish I Knew When I Was A Young Radio Artisan. I agree with most of his comments with the exception of this one:

Don't get your start on 2-meter repeaters.

This took me back to the comments from the Experienced Radio Amateur from years ago. I get the point — starting out on 2 meters can give you a limited view of ham radio — but I see it as the perfect platform for getting started. Here's what is working in my area with new Techs: get them started with a dual-band FM rig (usually an HT) so they have some on-the-air success. This also puts them in touch with the local ham community, where we not-so-subtlely expose them to other bands, modes and activities. They hear the other guys talking on the repeater about working DX on 10 Meters and start thinking about how to pursue that as a Tech. From there, it just expands out to all kinds of bands and modes.

Just for the record, I guess I did follow the advice of the Experienced Radio Amateur and managed to not "get stuck on two." That is, I've worked all of the bands from 160m to 10cm, earning WAS, WAC, DXCC and VUCC.

17. VHF Distance From Pikes Peak

I recently received this question via email from Dave/NØMUA:

> Bob, I ran across your pictures mountain topping on Colorado peaks, thought if anyone could answer this it would be you and your group. We run on 146.52 here in Coffeyville KS. and a group of us have brought up the question how far east can a mobile atop Pikes Peak be heard on 146.52 FM? The mobile would be mine running a Icom V8000 into a Tokyo HI power amp at 375 watts LMR 400 coax to a Cushcraft 13B2 beam pointing to the east.

This is one of those *how far will my signal go?* questions that always gets my attention. Other folks may find this interesting, so I decided to spend some time on the topic and post it here. I am assuming we are talking about tropospheric propagation and not something more exotic such as meteor scatter or sporadic-E skip.

The (incorrect) conventional wisdom is that VHF propagation is "line of sight", extending a bit beyond the optical horizon. From Wikipedia, we find that the distance to the optical horizon can be approximated by:

$$d = \sqrt{1.5h}$$

where *d* is the distance to the horizon in miles, *h* is the height of the observer above ground in feet

Pikes Peak reaches to 14,115 feet above sea level. The elevation of the surrounding area varies but since Dave is asking about propagation into Kansas, let's use a typical elevation of eastern Colorado (4500 feet). This gives us an optical horizon equal to the square root of (1.5 x 9615) = **120 miles**. Yes, this is an approximation, so feel free to knock yourself out with a more precise calculation.

It is interesting to note that there is a community 30 miles west of the Kansas border called Firstview, CO that is supposed to provide the first opportunity to see Pikes Peak when traveling from the east on Highway 40. Firstview is about 135 miles east of Pikes Peak, so the 120-mile calculation is in the right ballpark.

The ARRL website says that the radio horizon is about 15% longer than the optical horizon, so that means our line-of-sight radio horizon is about 1.15 x 120 = 138 miles. I've operated from the summit of Pikes during the ARRL June VHF Contest and the Colorado 14er Event and working stations on 2 meters at this distance is not difficult. To be more specific, I worked the Mt Sunflower crew (highest point in Kansas, 160 miles) from Pikes Peak on 2-meter FM using a 25 Watt mobile and a not-very-well-positioned quarter-wave antenna on the SUV fender.

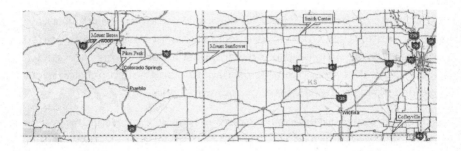

I pulled up the distance records for the Colorado 14er Event and found that the best DX using 2 Meter FM is when Phil/NØKE on Mount Bross worked Larry/NØLL in Smith Center, KS at a distance of 375 Miles! Clearly, we are well beyond line-of-sight for this radio contact. NØLL has a *very* capable VHF station on his end and I believe NØKE was using a decent Yagi antenna and was running some power(~200 W ?). Still, this contact was on FM which is not that great for weak signal work. While Pikes is on the Front Range of the Rockies, Mount Bross (14,172 feet) sits back some distance, about 60 miles west of Pikes Peak. (I have also worked NØLL from Pikes using 50W and a single 2M9 Yagi on SSB with no problem during VHF contests.)

Also during the Colorado 14er Event, Phil/NØKE (and Jeff/NØXDW) on Mount Bross worked W7XU in Parker, SD on 2-meter CW at a distance of 551 miles. Keep in mind that as the signal strength fades, SSB has a serious advantage over FM and CW is even better! So for squeaking out the marginal contacts, CW is the way to go.

Dave asked about working Pikes Peak from Coffeyville, KS using 2-meter FM. I had to look up where Coffeyville is and discovered that it is way the heck over on the east side of Kansas, maybe 50 miles from Missouri. I estimate that Coffeyville is 525 miles from Pikes Peak. To get back to Dave's question, making a contact from Pikes to Coffeyville on 146.52 MHz FM is not likely. Maybe if we got some exceptional tropospheric propagation...but I think even then it would be unlikely to complete the contact using FM.

But you never know what might happen on VHF. That's what makes it fun.

18. Yes, Band Plans Do Matter

There was an interesting exchange on the AMSAT-BB email list last week. Dave/KB5WIA noted a strange signal on the AO-51 satellite:

> I just thought I'd relay a bit of QRM I observed on AO-51 on this morning's 3/16/2011 1322z pass. The bird was totally quiet (just a nice carrier) for the first 5 minutes of the pass, but then it sounded like a repeater was getting into the sat uplink:
>
> 3/16 1327z: "Connected, KD7xxx repeater."
> 3/16 1328z: "KD7xxx repeater disconnected."
> 3/16 1328z: "hey Stacy did I see you at the corner there by Wendys?"
> 3/16 1330z: "...repeater in Middleton, Idaho."

I obscured the KD7 call sign to protect the ~~guilty~~ innocent. A little searching on the internet by some of the AMSAT folks revealed that there was an EchoLink station that matched the KD7 call sign.

Patrick/WD9EWK/VA7EWK wrote (again, I obscured the call signs):

> And there is an Echolink system (KD7xxx-R). What may be more interesting, after some Google searches, is a series of references I saw where the KC0xxx-L system had been linked to the KD7xxx-R system.
>
> On http://www.echolink.org/logins.jsp now (1839 UTC), I saw this for KC0xxx-L:
> KC0xxx-L Clay Cntr,KS 145.920 (1) ON 01:27 367513
> In the station description, it shows 145.920 along with the QTH in Kansas.
> This may be the system that's causing the QRM on AO-51, and the other system is just linked to KC0xxx-L at that time.

So it turns out that the KD7 call sign heard was linked to the KC0 EchoLink station which was operating on the uplink frequency of AO-51. George, KA3HSW, sent the KC0 operator an email and reported back that the KC0 station "has graciously changed frequencies."

What can we learn from this?

- Check the VHF band plans for your area before getting on the air. Be extra careful when setting up stations such as EchoLink or similar system that transmits frequently.
- Be aware that there are amateur radio modes that you can interfere with *even though you don't hear anything on frequency.* In the case of the AO-51 interference, the satellite hears the uplink frequency over a wide geography *but never transmits on that frequency.* The downlink is on the 70 cm band.
- Note that the first call sign associated with the interference (KD7xxx) was not at fault. It would have been easy to jump on his case and chew him out for transmitting on the satellite uplink frequency. Showing good judgment, the satellite guys investigated further.
- The issue was resolved by a polite (I assume) email to the offending radio amateur and he agreed to change the frequency of the EchoLink station. Nicely done.

So check the band plan for your area and follow it. And proceed with caution when interference does occur. It was a rookie error to put an EchoLink station in the satellite sub band and it was quickly resolved.

For more information on 2 Meter band usage, see Chapter 10: Choose your 2m Frequency Wisely.

19. Radio Hams are Not First Responders

It happened again. A disaster hits — this time a series of storms in the southeast— and the amateur radio community rises to the occasion to supply emergency communications.
See Tornadoes and Thunderstorms Keep Radio Amateurs Busy on the ARRL website.

I've noticed that there is a tendency for some members of the amateur radio community to characterize this activity as being a "first responder." (Most recently: Amateur Radio Newsline, 6 May 2011.) This may make for a more exciting story about how amateur radio operators assist during a disaster, but I think it is just sloppy terminology. Here's one definition of a *first responder* from dictionary.com:

> **first responder**
> –noun
> *a person who is certified to provide medical care in emergencies before more highly trained medical personnel arrive on the scene.*

The National Highway Traffic Safety Adminstration (NHTSA) has a 342-page standard that describes the training required to be a First Responder. Similarly, the wikipedia entry for Certified First Responder describes the skills necessary to be considered a First Responder. Most hams won't even come close to meeting this level of training, unless they happen to have it for reasons other than ham radio.

Why does this matter? By telling radio hams they are "first responders," it puts entirely the wrong emphasis on the Emergency Communications (EmComm) role. The EmComm role is, well, uh, providing *emergency communications*... in support of First Responders (Fire, Police, EMS) and agencies that support First Responders (Red Cross, Salvation Army, etc.). Where hams can really make a difference is making sure effective communications are established when a disaster occurs. The "I am a first responder" mindset can lead to some behavior that makes some radio amateurs look silly. The folks over at hamsexy.com have made a hobby out of ridiculing the so-called whackers that try to make ham radio into a lights-n-siren kind of operation. An even more serious issue is having radio hams engaging in dangerous activity without proper training.

Now, should radio hams get training on skills such as CPR and First Aid? Absolutely. Actually, everyone should have that training...you might find yourself in the situation of saving someone's life. But don't confuse that with being a trained First Responder.

> **Ham radio volunteers play an important role in providing emergency communications during disasters. This is usually in support of a served agency and not as first responders.**

20. Amateur Radio is Not for Talking

When the topic of ham radio comes up with normal people (that is, non-hams), I usually get asked the question "Who Do You Talk To?" I always come up with some vague and lame answer like "people all over the world" that gets me by.

I finally figured out what the problem is:

Amateur radio is not for talking to people.

At least not for me. If you just want to *talk to someone*, there are much better ways to do it, such as the telephone or Skype. There was a time long, long ago when one of the hooks for ham radio was "you can talk to a family member" when telephone service was not available or too expensive. Those days are gone. Back in the 1970s and 1980s, the hook was often "make mobile phone calls via the repeater autopatch." That was cool stuff that mere mortals could not do. With everyone and their dog now having a mobile phone, those days are gone, too.

So where does that leave us? Back where we started: the universal purpose of amateur radio is *to have fun messing around with radios*. Of course, this takes many different forms: public service, emergency communications, chasing DX, chasing counties, competing in contests, building kits, running QRP…to name just a few.

Now you will hear actual conversations on the ham bands but if you listen closely they usually have a radio underpinning to them. Hams are always talking about their equipment, signal strength, why their antenna is so great, why their antenna fell down, etc. It is kind of like making a telephone call where you talk about the quality of the phone line, the type of telephone being used and potential improvements to your home phone system.

Wait, you say, what about those guys on 75 Meters every evening talking about their medical conditions and complaining about the government? Aren't those guys actually talking to people? Those guys — I can't explain.

21. The History of Electronic Communications

Humans have always had a desire to communicate. They started out just talking to each other but then found that it was really handy to be able to write things down. This caused the invention of the alphabet and the training of English teachers to explain overly-complex rules of grammar.

In 1831, Joseph Henry was playing around with electric circuits and came up with the idea of a telegraph. We can imagine a simple system where ON means "time for dinner" and OFF means "not yet." This wasn't good enough for Samuel Morse, who invented the Morse Code which could use ON and OFF to represent the entire alphabet. This was the first digital code and was used for important messages such as "Laughing Out Loud", later abbreviated to LOL. This was basically the same as modern text messaging but you needed a trained telegrapher to do it.

In 1875, an inventor named Bell decided that it would be better if you could just *talk* over the wires instead of messing around with Morse Code. This will be a recurring theme — whether to talk to other people or just send digital codes. Bell invented this thing called the *telephone*, which is still used today. Basically, a person could talk into one end of a wire and have his voice pop out of the other end.

Later a guy named Marconi came along with the idea that communication should not depend on wires. For example, it was quite inconvenient to drag a telephone wire behind a ship as it moved across the sea. Unfortunately, Marconi didn't know how to do voice over the wireless, so he dropped back to using Morse Code. ON and OFF is a much easier way to go. Although there is a persistent rumor that Marconi intentionally used Morse Code to torment future generations of FCC Licensed Amateur Radio Operators, I can find no evidence of this. Some people argue that Nikola Tesla invented wireless but I think he had to be disqualified for overloading and shutting down the Colorado Springs power grid on numerous occasions.

Again, not wanting to be limited by trained telegraphers, voice communication (originally called *Amplitude Modulation*, but now known as *Ancient Modulation*) was invented. We are not sure who first came up with Ancient Modulation, but there are a bunch of radio hams on 75 Meters still trying to perfect it.

Somewhere around 1973, Motorola figured out that what the world really wanted was a portable phone that everyone could carry around in their pocket. The first attempt at this was the Motorola DynaTAC, which required an enormous pocket to carry it in. Knowing that customers were not going to enlarge their pockets, various mobile phone manufacturers worked feverishly to reduce the size of these phones.

Unfortunately, the mobile phone manufacturers terribly miscalculated, thinking that people would want to actually *talk on these phones*. As text messaging was added to these phones, it was discovered that most people, especially those under the age of 30, preferred to send cryptic text messages rather than actually speak to anyone. It was also discovered that all forms of human thought can be captured as 140 character messages. Although it was tempting to apply Morse Code to digital text messaging, it was rejected in favor of the ASCII 8-bit code. Instead of using ON/OFF keying, text messages are normally sent with a tiny keyboard patterned after a full-size typewriter (now obsolete).

By the way, I made up some of this stuff.

22. Seventy Three

Like many technical activities, amateur radio has its own set of jargon and protocols used both on and off the airwaves. As part of our Technician license course, we cover basic jargon but also encourage the use of plain language. A new Technician recently asked about the use of the term "73" on the local FM repeater, so I am posting this short piece.

Much of amateur radio history and practice is rooted in Morse Code, which traces back to the electrical telegraph. Two shorthand codes you'll hear on both voice and Morse Code communications are:

> **73 means *Best Regards***
> **88 means *Love and Kisses* (sometimes *Hugs and Kisses*)**

These codes originated with telegraph operating and are listed in the Western Union 92 Code, a set of numerical shorthand codes. On voice (phone) transmissions, you often hear something like this:

"Great to talk to you, Joe. Thanks and Seventy Three. This is KØNR, clear."

Since 73 is often used at the end of a radio contact, it almost takes on the meaning of "best regards and goodbye." "Eighty Eight" is used in a similar manner but is heard much less frequently on the ham bands.

Sometimes you'll hear 73 expressed as "Seven Three," which corresponds to how the Morse characters were sent. It is incorrect to say "Seventy-Three's" since this would literally mean "Best Regards's". Of course, most of us have made this error from time to time, very similar to grammatical errors in the English language. ("Somes time we forget to talk good.")

QRP operators often use 72 instead of 73 because low power operating is all about getting by with less. See W2LJ's blog.

And I normally use 73 at the end of most ham radio related email messages.

73, Bob KØNR

23. Proper Kerchunking

Recently, on one of the email reflectors associated with repeater owners, someone asked how to deal with kerchunkers on the repeater. The term *kerchunk* means to key up the repeater to see if it is there. It just takes a quick push of the Push-to-Talk (PTT) button on the transceiver to bring up most repeaters, resulting in a kerchunk sound.

It seems that this repeater owner had someone that was kerchunking his repeater on a regular basis and it was making him looney. This led to the usual discussion of whether kerchunking is acceptable, legal or moral and whether it should or should not be considered a capital offense.

Clearly, some radio amateurs have not been schooled in the proper way to kerchunk a repeater. The proper method for kerchunking is to key the transmitter and say your call sign, followed by the word "kerchunking." This simultaneously identifies your station and indicates the purpose of your transmission.

To make the practice of repeater kerchunking even more efficient, I am proposing the adoption of these new Q signals:

> QKC: *I am kerchunking the repeater*
> QKC?: *Are you kerchunking the repeater?*

Thank you for your attention to this important topic concerning good amateur practice.

Happy Kerchunking!

24. A Simple Wilderness Protocol

"The Wilderness Protocol" (ref. June 1996 QST, page 85), recommends that stations (fixed, portable or mobile) monitor the primary (and secondary if possible) frequency(s) every three hours starting at 7 AM local time, for five minutes (7:00-7:05 AM, 10:00-10:05 AM, etc.) Additionally, stations that have sufficient power resources should monitor for five minutes starting at the top of every hour, or even continuously." The primary frequency is the National Simplex Calling Frequency...146.52 MHz. The secondary frequencies are 446.0, 223.5, 52.525 and 1294.5 MHz.

Here in Colorado, the summer months mean that many people head for the mountains. Mobile phone coverage has improved in many parts of the high country but is still not reliable in all areas. Amateur radio VHF/UHF repeater coverage is extensive but also does not cover the entire state.

The Wilderness Protocol is a good idea but is overly complex for practical use. Here's my proposal to make it much simpler for practical backcountry use:

> Principle #1: Don't ever rely on a radio or mobile phone to get you out of trouble in the backcountry. Your primary strategy must be self-sufficiency. Avoid trouble. Be prepared for the unexpected.

Principle #2: Know what repeaters are available in your area. We have many wide-coverage repeaters available but you need to know the frequency, offset and CTCSS tone (if any). The Colorado Connection is a linked repeater system that covers many remote parts of the state.

Principle #3: In remote areas, monitor 146.52 MHz as much as possible. This applies to backcountry travelers, mobile stations and fixed stations.

I've been making it a habit to monitor 146.52 MHz in the backcountry. I often come across hikers, campers, fisherman, 4WD enthusiasts, SOTA stations, mobile operators and others monitoring that frequency. It is fun to chat with other radio amateurs having fun in the mountains.

25. Can I Use My Ham Radio on Public Safety Frequencies?

We have quite a few licensed radio amateurs that are members of public safety agencies, including fire departments, law enforcement agencies and search and rescue. Since they are authorized users of those public safety channels, they often ask this question:

> Can I use my VHF/UHF ham radio on the fire, police or SAR channel?

It is widely known that many amateur radios can be modified to transmit outside the ham bands. The answer to this question *used to be* that amateur radio equipment cannot be used legally on public safety channels because they are not approved for use under Part 90 of the FCC Rules. (Part 90 covers the Private Land Mobile Radio Services.) The only option was to buy a commercial radio with Part 90 approval and a frequency range that covered the desired amateur band. Some commercial radios tune easily to the adjacent ham band but some do not. The commercial gear is usually two to three times as expensive as the amateur gear, and just as important, does not have the features and controls that ham operators expect. Usually, the commercial radios do not have a VFO and are completely channelized, typically changeable only with the required programming software.

The situation has changed dramatically in the past few years. Several wireless manufacturers in China (Wouxun, Baofeng, Anytone, Tytera, etc.) have introduced low cost handheld transceivers into the US amateur market that are approved for Part 90 use. These radios offer keypad frequency entry and all of the usual features of a ham radio. It seems that these radios are a viable option for dual use: public safety and amateur radio, with some caveats.

> **Your transceiver must be approved by the FCC for Part 90 use if it is going to be used on public safety frequencies.**

New radios are being introduced frequently, so I won't try to list them here. However, you might want to do a search on Wouxun, Baofeng, Anytone and Tytera for the latest models. I will highlight the Wouxun KG-UV6D which I have been using. It seems to be a well-designed but still affordable (<$250) dual-band handheld radio.

Some Things to Consider When Buying These Radios

- The manufacturers sometimes offer several different radios under the same model number. Also, they are improving the radios every few months with firmware changes and feature updates. This causes confusion in the marketplace, so buy carefully.
- Make sure the vendor selling the radio indicates that the radio is approved for Part 90 use. I have seen some radios show up in the US without an FCC Part 90 label.
- Make sure the radio is specified to tune to the channels that you need.
- The 2.5-kHz tuning step is required for some public safety chan-

nels. For example, a 5-kHz frequency step can be used to select frequencies such as 155.1600 MHz and 154.2650 MHz. However, a 2.5 kHz step size is needed to select frequencies such as 155.7525 MHz. There are a number of Public Safety Interoperability Channels that require the 2.5-kHz step (e.g., VCALL10 155.7525 MHz, VCALL11 151.1375 MHz, VFIRE24 154.2725 MHz). The best thing to do for public safety use is to get a radio that tunes the 2.5-kHz steps.
- Many of these radios have two frequencies in the display, but only have one receiver, which scans back and forth between the two selected frequencies. This can be confusing when the radio locks onto a signal on one of the frequencies and ignores the other. Read the radio specifications carefully.

Recommendation

There are a number of reasonably good radios out there from various manufacturers. My favorite right now is the Wouxun KG-UV6D. The Anytone NSTIG-8R is a lower cost radio worth considering. The Baofeng UV-5R continues to be popular in the amateur community as the low cost leader. However if you show up at an incident with the Baofeng, your fellow first responders will probably think it is a toy.

Which leads to a really important point: the established commercial radio manufacturers such as Motorola, Vertex, etc. build very rugged radios. They are made for frequent, heavy use by people whose main job is putting out fires, rescuing people in trouble and dealing with criminals. These radios can take a lot of abuse and keep on going. The low-cost radios from China are not in the same league. Not even close. However, they can still serve in a less demanding physical environment while covering the Amateur Radio Service (FCC Part 97) and the Private Land Mobile Radio Services (FCC Part 90).

26. Phonetic Alphabets

For clear communications under all conditions, we use a phonetic alphabet for spelling out critical information. Instead of "A B C", we say "Alpha Bravo Charlie." Letters such as D, T and V can sound alike during noisy conditions, whereas Delta, Tango and Victor are more distinct. The standard phonetic alphabet for amateur radio comes from the International Telecommunication Union (ITU). This alphabet is also referred to as the NATO or International Aviation alphabet, although the spelling of the words may change slightly. This is the phonetic alphabet that you should commit to memory for ham radio use.

You will hear other phonetic alphabets used on the air from time to time. Also shown in the table is the "DX alphabet" and its alternate, which are popular on the HF bands for working DX and for contesting. Note that the DX alphabets tend to have more syllables which can sometimes be more effective. Sometimes it is helpful to switch to a different phonetic alphabet to punch through. Sometimes that gets the message through.

Because of these variations, you may think it's OK to make up your own phonetics. Some hams like to come up with something cute and easy to remember for their own call sign. A call sign such as W2LPR might be "Whiskey Two Long Playing Record." Certainly easy to remember but if you use these phonetics on the air under marginal conditions, you'll probably just confuse the operator on the other end.

Adapted from KC4GZX

	ITU	DX	DX Alternative
A	**Alpha**	America	Amsterdam
B	**Bravo**	Boston	Baltimore
C	**Charlie**	Canada	Chile
D	**Delta**	Denmark	
E	**Echo**	England	Egypt
F	**Foxtrot**	France	Finland
G	**Golf**	Germany	Geneva
H	**Hotel**	Honolulu	Hawaii
I	**India**	Italy	Italy
J	**Juliet**	Japan	Japan
K	**Kilo**	Kilowatt	Kentucky
L	**Lima**	London	Luxembourg
M	**Mike**	Mexico	Montreal
N	**November**	Norway	Nicaragua
O	**Oscar**	Ontario	Ocean
P	**Papa**	Pacific	Portugal
Q	**Quebec**	Quebec	Queen
R	**Romeo**	Radio	Romania
S	**Sierra**	Santiago	Sweden
T	**Tango**	Tokyo	Texas
U	**Uniform**	United	Uruguay
V	**Victor**	Victoria	Venezuela
W	**Whiskey**	Washington	
X	**X-Ray**	X-Ray	
Y	**Yankee**	Yokohama	
Z	**Zulu**	Zanzibar	Zulu

Most of the time, I stick to the ITU phonetics but I may use the DX phonetics for contests. The ITU phonetics for my call sign are "Kilo Zero November Romeo," but I'll often switch to "Kilo Zero Norway Radio," which is a few syllables shorter. If the other operator is having trouble picking my call sign out of the noise, it sometimes helps to switch phonetic alphabets. For example, if someone is struggling to understand the "R" in my call sign when I say "Romeo", I may switch to "Radio" to try something different. Sometimes one or the other sound just happens to get through better or is more recognizable by the other radio operator (especially if English is not their primary language).

Some people will tell you that you don't need to use phonetics on VHF FM because the signals are so clear. I've even heard hams say that its *wrong* to use phonetics on FM. This is clearly short-sighted because we often have noisy and marginal conditions when using FM. Vehicle noise, power line noise and weak-signal conditions can cause us to struggle with hearing the other operator clearly. When this happens, go ahead and use phonetics.

References

KC4GZX Phonetic Alphabet Page
https://www.qsl.net/k/kc4gzx/kc4gzx/chrtphonetic.htm

27. Twisted Phonetic Alphabet

I recently posted an article about the use of phonetic alphabets. See Chapter 26: Phonetic Alphabets. The "standard" phonetic alphabet is the ITU alphabet but I am starting to think that we might need to get a little more creative on our use of phonetics. Why not innovate in this area, just like we innovate on the technical front?

Towards that end, I was reminded of his phonetic alphabet listed over at netfunny.com:

A	Are		N	Nine
B	Bee		O	Owe
C	Cite		P	Pseudonym
D	Double-U		Q	Queue
E	Eye		R	Rap
F	Five		S	Sea
G	Genre		T	Tsunami
H	Hoe		U	Understand?
I	I		V	Vie
J	Junta		W	Why
K	Knot		X	Xylophone
L	Lye		Y	You
M	Me		Z	Zero

Even this creative alphabet can be improved on. For example, I think H should be Honor.

28. Go Ahead and Use Phonetics on 2m FM

Sometimes radio amateurs suggest that phonetics are not needed on VHF FM. Sometimes it even sounds like it's a bad thing to use phonetics on FM. *It is inefficient and slows things down.* I can see the logic behind this because with decent signal strength, demodulated FM audio is usually quite clear and easy to understand.

Here's what I wrote in Chapter 3: **VHF FM Operating Guide**, also downplaying the need for phonetics:

> The use of phonetics is not usually required due to the clear audio normally associated with frequency modulation. Still, sometimes it is difficult to tell the difference between similar sounding letters such as "P" and "B". Under such conditions, use the standard ITU phonetics to maintain clarity. Many nets specifically request the use of standard phonetics to make it easier on the net control station.

The FCC Technician exam gives the topic of phonetics a light touch with just these two questions:

T1A03 (D)
What are the FCC rules regarding the use of a phonetic alphabet for station identification in the Amateur Radio Service?

A. It is required when transmitting emergency messages
B. It is prohibited
C. It is required when in contact with foreign stations
D. It is encouraged

And this one:

T2C03 (C)
What should be done when using voice modes to ensure that voice messages containing unusual words are received correctly?

A. Send the words by voice and Morse code
B. Speak very loudly into the microphone
C. Spell the words using a standard phonetic alphabet
D. All of these choices are correct

Use Phonetics

In practical radio operating, there are a number of things that can degrade communication, usually by creating noise sources that compete with the voice modulation. Most of these are a factor *even if the RF signal is strong:*

- A noisy environment at the receiving end (e.g., background noise such as road noise in an automobile)
- A noisy environment at the transmitter (e.g., background noise such as wind noise outdoors)
- Poor frequency response of the overall system (e.g., high frequencies may be lost in the transmitter, receiver or repeater, making it more difficult to understand the voice communication).
- Hearing impairment of the person receiving the audio (I've heard that we are all getting older)
- Difficulty understanding the person speaking (poor enunciation, unfamiliar dialect or accent, etc.)

So I say go ahead and use phonetics on VHF FM, especially for critical information such as your call sign. FM communication is *not* always clear and easy to understand. It suffers from the same signal-to-noise problems as other voice modes. (Perhaps not as bad as SSB on HF, but it's still a factor.) In most cases, you'll want to stick with the standard ITU phonetic alphabet (also known as the NATO alphabet).

The ITU Phonetics are

A – Alpha	N – November
B – Bravo	O – Oscar
C – Charlie	P – Papa
D – Delta	Q – Quebec
E – Echo	R – Romeo
F – Foxtrot	S – Sierra
G – Golf	T – Tango
H – Hotel	U – Uniform
I – India	V – Victor
J – Juliet	W – Whiskey
K – Kilo	X – X-ray
L – Lima	Y – Yankee
M – Mike	Z – Zulu

Many nets request that you use ITU phonetics when you check in. Imagine being the Net Control Station for a net and having everyone making up their own phonetics. You would have call signs coming at you with all kinds of random words associated with them. It is much better to have consistency. However, there are times when you might want to use alternative phonetics.

73, Bob
Kilo Zero November Romeo

29. Go Ahead and Call CQ on 2m FM

The conventional wisdom in amateur radio is that we should not call CQ when using FM on the VHF and UHF bands, especially on repeaters. The reasoning for this is that during normal VHF/UHF FM operating, radio amateurs are tuned to a specific frequency and will easily hear a call on FM. Compare this to the HF bands, where the other ham is generally tuning around to find someone to contact and stumbles onto your transmission. In that case, you want to make a long call (CQ CQ CQ Hello CQ This is Kilo Zero November Romeo calling CQ CQ CQ...) so people tuning the band will find you and tune you in. On VHF/UHF FM, the assumption is that the other hams have their radio set on the repeater or simplex channel being used and will immediately hear you. FM communications are often quite clear and noise free, which also helps. The normal calling method is to just say your call sign, perhaps accompanied with another word like "monitoring" or "listening." For example, I might say "KØNR monitoring."

Question T2A09 in Technician exam pool reinforces this idea:

> T2A09 (B) What brief statement indicates that you are listening on a repeater and looking for a contact?
> A. The words "Hello test" followed by your call sign
> **B. Your call sign**
> C. The repeater call sign followed by your call sign
> D. The letters "QSY" followed by your call sign

Gary/KN4AQ wrote this tongue-in-cheek article HamRadioNow: Do NOT Call CQ on Repeaters which says that calling CQ on a quiet repeater works well because it is likely that someone will come on and tell you not to call CQ. Gary wrote:

> So I trot out my standard advice: **make some noise.** I even recommend **calling CQ**, because that's almost guaranteed to get someone to respond, if only to tell you that **you're not supposed to call CQ on repeaters.**

Scanning and Multitasking

Some important things have changed in our use of VHF/UHF FM over past decades. The most important shift is dispersion of activity: while the number of VHF/UHF channels has increased, the total amount of VHF/UHF radio activity has declined. This means that we have tons of channels available that are mostly quiet. Tune the bands above 50 MHz and you'll hear a lot of dead air. In response to this, some hams routinely scan multiple repeater and simplex frequencies. While getting ready for Summits On The Air (SOTA) activity, I've had hams ask me to make a long call on 146.52 MHz so they can be sure to pick me up on scan. Another factor that comes into play is the multitasking nature of our society. Hams don't generally sit in front of a 2-meter radio waiting for activity to occur. More commonly, they are doing something else and listening to the FM rig in the background. VHF FM is the Utility Mode, always available but not necessarily the top priority. A short call ("KØNR listening") on the frequency can easily be missed.

Recommendations

My conclusion is that the Old School "KØNR Monitoring" style of making a call on VHF is no longer sufficient. First off, it sends the message of "I am here if you want to talk to me." If that's your intent, fine. However, if you really want to make a contact, being more explicit and a bit assertive usually helps. Follow Gary's advice and *make some noise*. For example, during a SOTA activation I'll usually call on 146.52 MHz with a bit of a sales pitch. Something like: "CQ CQ 2 meters, this is Kilo Zero November Romeo on Pikes Peak, Summits On The Air, anyone around?" This is way more effective than "KØNR Monitoring." I might also include the frequency that I am calling on, to help out those Scanning Hams. Something like "CQ CQ 146.52, this is KØNR on Pikes Peak, Summits On The Air." Note that these calls are still pretty much short and to the point, only taking about 15 seconds. This is a lot shorter than the typical HF CQ. If I am driving through another town and want to make contact on the local repeater, I will adjust my approach accordingly. For example, on a relatively quiet repeater, I might say "CQ, anyone around this morning? KØNR mobile I-25 Denver." Or if I have a specific need, I'll go ahead and ask for it. "This is KØNR looking for a signal report." Keep in mind that VHF/UHF operating tends to be local in nature, so it makes sense to adapt your approach to both local practice and the specific situation.

- **It's OK to call CQ on VHF FM, make some noise on the frequency.**
- **Give other operators a reason to contact you.**
- **Don't make your call too long, maybe 15 to 20 seconds.**
- **The call sign/monitoring approach is fine, too.**

30. Religion and Ham Radio

We need to get the religion out of ham radio. No, I am not talking about the HF nets that support missionaries or similar activities. (Those people might actually be doing something good for the world.) I am talking about the religious debates concerning new technology...*this* technology is better than *that* technology.

Amateur radio is a technical hobby, one based on technology, hobbyist pursuits and mutual interest. One might think that this means issues are looked at objectively and discussions are based on logic, scientific principles and facts. Of course, this is completely wrong. What often shows up in ham radio are *religious* debates about technology or operating modes.

Here's a definition of Religion:

> a specific fundamental set of beliefs and practices generally agreed upon by a number of persons or sects: *the Christian religion; the Buddhist religion.*

You can tell when you are stuck in a religious debate...the facts quickly fade and statements like "this is the *right way* to do it" become louder. Underneath this is a fundamental belief trying to come out that the person may not even be aware they have.

A long running example of a religious debate is *Linux* versus *Windows*. On the surface, people argue about which one has more defects, which one is more secure and which one ultimately serves their needs better. Underneath the surface is the religious belief: *software should be free, Microsoft is evil, etc*. Then there are those Mac enthusiasts (you know who you are)....these folks tend to act like a cult as they attempt to convert other people to their group. (Where is the line between *enthusiast* and *cult member?*)

The latest one on the ham radio front is the debate over digital technology in the VHF and higher bands: D-STAR versus DMR (and now Fusion). The debate starts out rational with a discussion of the merits of each but soon the deeply-held beliefs come out: *D-STAR is bad because ICOM is pushing it, DMR is good because it is the commercial standard, D-STAR is good because it is an amateur radio standard, D-STAR uses a proprietary vocoder chip so it is bad, etc*. Then don't forget the guys that say "all digital is bad, analog FM is good."

Again, you can tell when the religion kicks in because the facts start to fade and the beliefs rise to the surface. Usually, these arguments can't be resolved because you can't really debate beliefs. What you get instead are flame wars on the various email groups.

What other religious debates are out there? *Android* versus *iOS*, *Open Source Software* versus *Commercial Software*, *My favorite rig* versus *Your favorite rig*, ... what else?

31. Sorry, I've Been On 2m FM Again

I was looking out the window the other day and noticed that my wire HF antenna is laying on the ground. Hmmm, probably doesn't radiate very well that way. But if I put a long, lossy coaxial cable in line, the SWR will still be good at the transmitter. And I can tell my buddies that it works just fine because "I can work everyone that I hear." (What a dumb thing to say.)

This made me realize that most of my ham radio activity lately has been on 2m FM. Actually it has been on 2m and 70cm FM, as I tend to lump these two activities together. These days, my VHF/UHF FM rigs have at least 146 MHz and 440 MHz in them (FT-7800, FT-8900, etc.). I cruise down the road and flip on the rig, talk to the locals, talk to my wife, etc. It is just too easy and too convenient. It fits the mobile lifestyle, whether it means operating a mobile rig in the car or grabbing an HT to take along on a business trip. (I used to run HF and SSB VHF mobile but found that the rigs were rarely used, so I removed the gear from my vehicle.)

Of course, I need to apologize to the rest of the ham community for this failure to act according to accepted social norms. You know how it is...Real Hams operate HF, weak-signal VHF, microwaves, etc......almost anything that is not 2m FM. Every so often I hear that comment about "well, those techs just hang out on 2m FM," implying that those guys are permanently stuck in ham radio middle school, unable to graduate to the next level. Or sometimes the FM operators are referred to as having "shacks on the belt" which are dependent on the "box on the hill." The main message is that 2m FM is just too easy, too plug-n-play, too much like an appliance….too convenient. We certainly can't have that!

Don't get me wrong…I enjoy HF, DXing, contesting, digital modes, almost anything to do with amateur radio. That's the cool thing about the hobby…so many bands, so many modes. One of my favorite activities is operating the major VHF contests. (I've even been known to make a few CW contacts.) But on a day-to-day basis 2m FM just seems to fit in better.

Some people call 2m FM *the Utility Mode*, because it is the mode that gets the job done. Last week, we had a weather net activated to track thunderstorms and a few tornadoes. Did this happen on 40m? I don't think so. Two meters carried the load. Where do most of the ARES and RACES nets meet? Two meters. How are most public service communications handled? Two meter FM. Even some hard core HF DX enthusiasts are known to flip over to 2m FM to tell their buddies that the DXpedition to a rare country is on the air. It is the *Utility Mode*.

Over the weekend, I was driving through the mountains and heard an aeronautical mobile working stations simplex on 146.52 MHz…lots of fun. Another time, I heard a station calling about 80 miles away (I was in a high spot) and I had the pleasure of making that contact….again, on 2m FM. A few weeks ago, I operated in the Colorado 14er Event from the summit of Pikes Peak. Since many of the mountaintop stations had hiked up, the most popular mode of the day was (you guessed it) 2m FM.

So sorry, I have been hanging out on 2m FM. I'll try to get that HF antenna back in the air one of these days.

32. What's In Your Rubber Duck?

Anyone with a VHF or UHF handheld transceiver (HT) probably uses the standard "rubber duck" antenna for casual use. I often refer to the rubber duck as *The World's Most Convenient Crappy Antenna*. To be fair, all antennas are a compromise…the rubber duck optimizes small size and convenience at the expense of performance. The Wikipedia entry describes the rubber duck antenna as "an electrically short monopole antenna…[that] consists of a springy wire in the shape of a narrow helix, sealed in a rubber or plastic jacket to protect the antenna."

Being curious about what really is hiding inside the typical rubber duck antenna, I decided to take a few of them apart. I did not try to assess the performance of the antennas but just examine their construction.

Baofeng UV-5R Stock Antenna

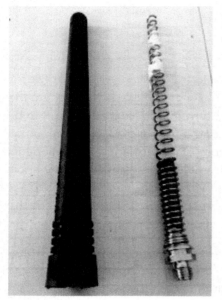

Baofeng UV-5R Antenna

I started by dissecting a Baofeng UV-5R antenna, which took some aggressive action with a diagonal wire cutters to split the rubberized jacket near the bottom. After that, the jacket slid off to reveal the classic spiral antenna element inside. You can see some white adhesive near the top of the spiral element (upper right in the photo). The Baofeng antenna had a female SMA connector.

Yaesu FT-1DR Stock Antenna

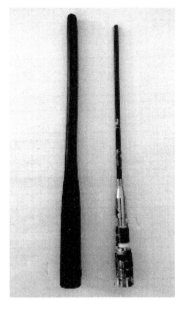

Yaesu FT-1DR antenna

The Yaesu antenna was easy to disassemble. In fact, I chose this antenna because I noticed that the outside jacket had come loose and was starting to slide off the antenna. A steady pull on the cover exposed the antenna elements without any further antenna abuse. (I plan to reinstall the cover with a few dabs of glue and expect that it will continue to work fine.)

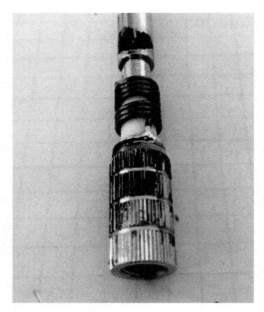

Close up of the FT-1D antenna coil

The construction of this antenna is quite different from the Baofeng. The main element is a very tightly-wound spring…so tight that I expect that it acts like a solid wire electrically. In other words, it doesn't have the spiral configuration that makes the antenna act longer electrically. At the bottom of the antenna, there is a coil inserted in series with the radiating element (connects radiating element with the center pin of the SMA connector).

Laird VHF Antenna

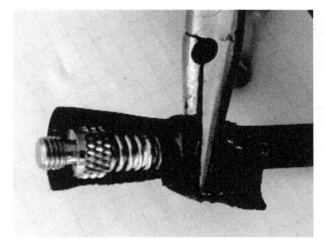

Peeling back the outer coating of the Laird antenna

Next, I wondered if antennas for commercial radios used different design or construction techniques. Laird makes high-quality antennas for the mobile radio and other commercial markets, so I purchased one of their VHF rubber duck antennas to dissect. This model is intended for use with Motorola radios requiring a threaded antenna stud.

This antenna was a challenge to cut open. I used a sharp knife and diagonal pliers to cut the rubberized jacket and peeled it back using a needle-nose pliers. The rubberized coating was embedded into the spiral antenna element, so it did not come apart easily. It took over an hour fighting with the antenna and I gave up before getting the entire spiral element exposed.

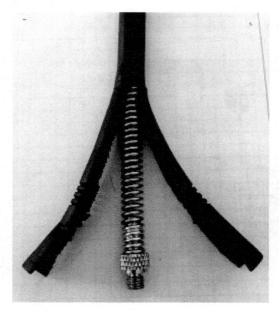

Laird VHF antenna with rubberized coating removed.

The Laird antenna is clearly the sturdiest of the three antennas. The spiral element is much thicker than the Baofeng and the rubberized coating is tougher and molded tightly into the spiral element. The Baofeng and Laird antennas use the same design concept...just take a spiral antenna element and apply a protective cover. However, the Laird construction was far superior, but not a surprise given that Baofeng is a low-cost provider in the ham radio (consumer) market. My disappointment is with the Yaesu antenna. The antenna came apart after one year of not very heavy use. I expect I can put it back together with some adhesive, improving on the design in the process.

What's in your rubber duck?

References

Rubber Ducky Antenna (Wikipedia)
https://en.wikipedia.org/wiki/Rubber_ducky_antenna

33. A Better Antenna for Dual-Band Handhelds

I'm a fan of using a half-wave antenna on a 2m handheld transceiver (HT). These come in a variety of forms but I've tended to use the telescoping half-waves that mount on the HT. These include the Halfwave 2 Meter Flex antenna from Smiley and the MFJ-1714 from MFJ. One of the disadvantages of these two antennas is that they are designed for 2m operation only. Put it on a dual-band HT and you can only use one of the bands.

The RH 770 dual-band antenna

Now there is a dual-band alternative. During a discussion of various VHF radios and antennas on the SOTA reflector, Phil/G4OBK recommended this antenna: TWAYRDIO RH 770 SMA-Male Dual Band Telescopic Handheld Antenna. I was mildly skeptical in that the antenna looks like ~~cheap~~ ~~lowcost~~ economy stuff from China. However, for $16.55 (free shipping), it seemed like something I should try out.

I've since used this antenna on several SOTA activations and have found it to work quite well. Not having to worry about whether I'm operating on 2 m or 70 cm is a big plus. I liked the antenna so much, I now have three.

Recently, I wondered how well the antenna is really performing so I did a side-by-side comparison with the Smiley 2m half-wave. Now this kind of comparison is always a bit dicey unless you have a calibrated antenna range but simple comparisons are useful. I got on 2m FM with another ham running a home station some distance away such that I was not pegging his S-meter. We did several A/B comparisons between the Smiley and the RH 770. Much to my surprise, the RH 770 performed significantly better than the Smiley. That is, the other ham saw his meter deflect higher with the RH 770. I can't give that to you in dB but I can say it's a little better. I actually thought that the single-band design might win out due to less complexity in the antenna but the opposite was true. Your mileage may vary. No warranty expressed or implied.

The only thing I don't like about the RH 770 is that the telescoping sections slide up and down really easy. Too easy for my taste. I'd rather have some stickiness to it so that I am sure it will remain fully extended. But I admit this is more of a personal impression than actual problem.

The antenna is available with a male SMA connector, a female SMA connector or a BNC. That should pretty much cover it.

So thanks Phil/G4OBK for pointing out this antenna. I also highly recommend it.

34. Is the Internet Destroying Amateur Radio?

How many times do you hear the comment "ham radio...do people still do that?" followed by the statement that "surely the internet has made ham radio obsolete." For the most part, that misses the point about the use and attractiveness of amateur radio.

I've written before that Amateur Radio Is Not for Talking and that the universal purpose of ham radio is *to have fun messing around with radios.* One significant statistic is that the number of FCC amateur radio licensees remain at an all time high. Eventually, the demographics will likely catch up with us and this number will start to decline, but it hasn't happened yet.

The internet has become a tool that is used to complement amateur radio, often in ways that we may not have predicted. Although there are plenty of "keep the internet out of amateur radio" folks in the hobby, there are many more that have found clever ways to make use of the internet. I view emerging technologies and technological innovation as unstoppable forces that will impact us whether we try to ignore them or not. Using that lens, let's examine the impact of the internet on amateur radio.

Here are a few broad categories of impact:

1. Communication Pipe

The internet is often used to provide an additional mechanism for transporting ham radio communications. Obvious examples are VoIP systems such as EchoLink and IRLP. Also included in this category are digital voice systems that use the internet to connect radios together: D-STAR, Yaesu System Fusion, Brandmeister Network, DMR-MARC Network. WinLink is a global email system using ham radio. The core transport technology is the Internet Protocol Suite (TCP/IP) which is not limited to the public internet. Some ham radio organizations are implementing IP links using microwave gear on the amateur radio bands so they are independent of the internet.

Another application in this category is remote operation of ham stations. That is, use an internet connection to control a ham station at another location. Sometimes people refer to this as the *Long Microphone Cord Model* (or maybe I just made that up). Hams do this with their own private stations but there are also shared stations established by radio clubs and commercial vendors (see Remote Ham Radio). With community restrictions on external antennas being very common, having a remote station available is very attractive.

This has turned out to be quite disruptive because so much of ham radio operating depends on your location, which is generally determined by the location of the transmitter. But now you can have a person sitting in downtown Denver operating a transmitter that is in Fiji. Kind of confuses things a bit. Regulatory issues also come into play: that transmitter in Fiji is going to fall under Fiji regulation which usually means needing an amateur radio license issued by the local government. The day is coming when a DXpedition to a remote island will consist of a helicopter delivery of a remote radio box (with satellite link and self-deploying HF antenna) that is operated by someone sitting at home using their smartphone.

2. Reporting and Coordination

Ham radio operators also use the internet for spotting and reporting purposes. *Spotting* has been around for a long time, which basically means letting other hams know that a particular station is on the air and can be worked from a particular location. Hams have done this without the internet but the internet certainly allows for more efficiency, or at least a lot more spots. DX Maps is a good example of a spotting website that supports lists and mapping of spots.

Radio hams also use the internet for coordinating radio contacts. One of the most extreme examples is the use of pingjockey for arranging meteor scatter communications. Typically, two hams will connect on pingjockey and agree to try a meteor contact on a specific frequency, with specific timing, etc. This technique is easy to abuse, either intentionally or via sloppy operating habits, because you can inadvertently share the radio contact information via the internet. However, properly used, pingjockey is a wonderful tool that promotes meteor scatter operating. ON4KST operates an amateur radio chat website that enables a wide variety of online communication and coordination between hams.

The Reverse Beacon Network (RBN) is a network of radio receivers listening to the amateur bands and reporting what stations they hear. These stations are often referred to as *CW Skimmers* because they skim the CW information from the received signals. RBN began with decoding CW but now also supports RTTY. There's no fundamental reason it couldn't be extended to other modes, even voice modes, with sufficient computing power.

PSK Reporter is a similar reporting system which accumulates signal reports from HF digital stations. As the name implies, it was first focused on PSK31 but has expanded to include other digital modes. Weak Signal Propagation Reporter (WSPR) is a more advanced propagation reporting system that uses transceivers and advanced DSP techniques. The compressed protocol sends the transmitting station's call sign, Maidenhead grid locator, and transmitter power in dBm. WSPR lights up the world with low power transmitters and measures HF propagation on all bands in real time. Very clever system.

These worldwide networks produce a very complete picture of available propagation and stations on the air. Some hams complain that "nobody tunes the dial" anymore because they just rely on the station of interest to be spotted. DX stations often have the experience of huge pileup as soon as they are spotted on one of the networks.

3. Logging and Confirmations

For decades, hams have been keeping their radio logs using a wide range of software that is available. This is a handy way of keeping track of radio contacts and tracking progress towards operating awards. More recently, online systems have been developed to allow radio contacts to be confirmed electronically. That is, instead of exchanging QSL cards as confirmation of a radio contact, both hams submit their log information to a central server that records the radio contact. The ARRL offers the Logbook of The World (LoTW) which supports these awards: DXCC, WAS, VUCC and CQ WPX. The eQSL website was the first online QSL site, offering electronic QSL card delivery and its own set of operating awards. Club Log is another online electronic logging system. The popular qrz.com website has added a logbook feature to its set of features.

Electronic confirmation of radio contacts is a huge improvement for ham radio. While many of us still enjoy getting a paper QSL card, collecting QSLs for awards is a royal pain. Mailing QSL cards is expensive, takes time and often involves long delays.

Impact on Amateur Radio

Here's my analysis of the situation: Categories 2 and 3 mostly represent a net positive influence on amateur radio. These are straight up information age applications that provide useful and quick updates about radio propagation and radio contacts. Yes, there is some downside in that many hams become dependent on them instead of doing it the old fashioned way: *turn the big knob on the radio and listen.* Not a big deal given the benefits.

Category 1 is more of an issue for me. The major effect is that it makes ham radio worldwide communication a lot easier. This is what causes many hams to say "That's Not Real Ham Radio" when the internet is used to do so much of the work. Focusing on the actual radio wave propagation, there is really no comparison between working DX on the 15m band and making the same QSO with a UHF DMR handheld piped through the internet. At this point, *I try not to overthink* the issue, dropping back to "have fun messing around with radios." So if chasing DX on 15m floats your boat, keep on doing it. If the DMR handheld provides enjoyment for you, I'm OK with that, too.

Perhaps more importantly, we can't really stop the impact of new technology. Oh, I suppose the amateur radio community could petition the FCC to restrict Category 1 use of ham radio. There could be regulations that limit the use of the internet being interconnected with Part 97 radio operation. However, that would have an even bigger negative impact on the hobby by arbitrarily restricting innovation. Imagine if we had to tell technically-minded newbies in the hobby that "well, we have this rule that says you can't actually use the biggest technology shift in the 21st century" while using ham radio. We do have some rules concerning awards and contests such as you can't use a VoIP network to qualify for DXCC. There will probably be more of that kind of restriction occurring as technology moves forward, which is fine by me.

What's Next?

When it comes to technological change, it's often difficult to predict the future. Some of it is obvious: we'll see higher bandwidths and more wireless coverage on the planet as 5G and other future technologies roll out. Figuring out how this affects ham radio is a bit more difficult. Right now, there are still remote locations that aren't on the network but that will change. I expect even remote DXpeditions to eventually have excellent connectivity which could lead to *instant check QSLs*. (That's kind of happening already but it could become more of a realtime event.)

As systems become smarter (e.g., machine learning, artificial intelligence), distributed systems will become more automated. We can expect more automation of ham radio activity which will certainly be controversial. Did you really work that other station if the software in your home ham station made the contact while you were away at work?

To wrap up, I don't think the internet is destroying amateur radio but it is certainly changing it. The key is to keep having fun and enjoying the hobby. If you aren't having fun, you probably aren't doing it right.

References

Remote Ham Radio – http://www.remotehamradio.com/

Logbook of The World (LoTW) – https://lotw.arrl.org/

35. That's Not Real Ham Radio

Things had been pretty quiet on the ham front lately but then I ran into a string of "That's Not Real Ham Radio" discussions. This happens from time to time...I usually ignore it...but this time I got sucked into the topic.

It started with some HF enthusiasts I know talking about how "digital modes" are just not very satisfying. Their point is that with CW and SSB, there is an audio connection to your ear that makes you an integral part of the radio communication. The extreme-DSP modes (WSJT) insert serious signal processing that essentially removes the human connection. This can quickly lead to the generalization that these digital modes "aren't real ham radio."

I think its fair to say that most hams think of the HF bands as the center of the hobby...getting on the air, bouncing signals off the ionosphere to talk to someone over the horizon. Some hams will go even further and say that CW is the only way to go. *Anything less is just phone*. FM and repeaters? Forget that stuff...not enough skill required. And certainly, don't get stuck on 2 meters.

In a previous post, I argued we should not confuse religion with modulation. I do occasionally make snarky comments about the continued use of AM (AKA Ancient Modulation), but I've tried to tone that down.

What About DMR?

Just last week, I was playing around with a DMR hotspot on the Brandmeister network. It really struck me that people on the system were having a blast talking to each other across North America and around the world. But then that nagging little voice in the back of my head said "hey, wait a minute...this is not real DX...the RF signal might only be traveling 20 feet or so from an HT to a hotspot."

This caused me to put out a plea for insight on twitter:

BobW @K0NR · Mar 12

Serious question: what is the attraction of a ham radio contact using RF to go 25 feet to a hotspot then around the world via TCP/IP?

I received a lot of good replies with the answers tending to clump into these three categories:

- I don't know ("That's Not Real Ham Radio")
- It's fun, new technology
- It's a digital network that brings ham radio operators together

My interest seems to fall into the second category: *this is fun, new technology.* Which does make me wonder how long this new technology will remain interesting to me. Well, that is difficult to predict but I'll invoke the principle of *try not to overthink it.* The idea that DMR is a *digital network that brings ham radio operators together* makes some sense. In the past, I have argued that amateur radio is not for talking. In other words, if you just want to talk to someone, there are much more convenient ways of doing that. Still, there is something attractive about this ham-radio-only digital network.

It really is important to not overthink this kind of stuff. Ham radio is supposed to be fun, so if you are having fun, you are probably doing it right. If you are not having fun, then you might want to examine what you are doing.

Sometimes hams can get a little spun up about those other guys that don't appreciate our way of doing ham radio. What the heck is wrong with them anyway? I've always been inspired by the Noise Blankers Mission Statement:

> **Do radio stuff.**
> **Have fun doing it.**
> **Show people just how fun it is.**

If your preferred form of ham radio is so superior, it ought to be easy to show other hams how cool it is. If not, then maybe you aren't doing it right. Conversely, as long as other hams are having fun and operating legally, don't knock what they are doing. In fact, encourage them. We need more people having fun with ham radio, even if it's not your favorite kind of fun.

36. Amateur Radio: Narrowband Communications in a Broadband World

For my day job in the test and measurement industry, I get involved in measurement solutions for wireless communications. Right now, the big technology wave that is about to hit is known as 5G (fifth-generation wireless). Your mobile phone probably does 4G or LTE as well as the older 3G digital mobile standard. For more detail on LTE, see ExtremeTech explains: What is LTE?

5G will be the next cool thing with early rollouts happening now and more planned in the coming years. The design goals of 5G are very aggressive, with maximum download speeds of up to 20 gigabits per second. (See what I did there: I used the words "up to", so don't expect this performance under all conditions.) The actual user experience has yet to play out but we can assume that 5G is going to be blazing fast. For more details see: Everything You Need to Know About 5G.

To achieve these high bandwidths, 5G will use spectrum at higher frequencies. Move up in frequency and you inherently get more spectrum. The FCC recently allocated 11 GHz of new spectrum for 5G, including allocations at 28 GHz, 37 GHz, 39 GHz and 64 to 71 GHz. Yes, those frequencies are GHz with a G...that's a lot of cycles per second.

Amateur Radio

So my day job is focused on wider bandwidths and higher frequencies. Then I go home and play amateur radio which is a narrowband, low frequency activity. The heart of ham radio operation is on the HF bands, 3 to 30 MHz, almost DC by 5G standards. Many of us enjoy VHF and UHF but even then most of the activity is centered on 50 MHz, 144 MHz, maybe 432 MHz. I recently started using 1.2 GHz for Summits On The Air, so that at least gets me into the GHz-with-a-G category.

Not only does ham radio stay on the low end of the frequency range, we also use low bandwidth. The typical phone emission on the HF bands is a 3-kHz wide SSB signal. That's kHz with a k. As we go higher in frequency, some of our signals are "wideband" such as a 16-kHz wide FM signal on the 2-meter band. In terms of digital modes, AX.25 packet radio and APRS typically use 1200 baud data rates but sometimes we go with a "super-fast" 9600 baud transmission mode. (Not really.)

CW is still a very popular narrowband mode with bandwidths around 200 Hz, depending on Morse code operating speed. Lately, the trend has been to go even narrower in bandwidth to keep the noise out and operate at amazingly low signal-to-noise ratios. Some of the WSJT modes use bandwidths in the range of 4 to 50 Hz.

There are some good reasons that amateur radio remains narrowband. The two most important are:

1. We love the ionosphere and what it does for radio propagation. The HF bands are great for making radio signals go around the world but they are narrow spectrum. For example, the 20-meter

band is 350 kHz wide, going from 14.000 to 14.350 MHz. Operation is restricted to narrowband modes, else we'd use up the entire band with just a few signals.
2. We just want to make the contact (and maybe talk a bit). For the most part, radio hams are just trying to *make the contact*. This is most pronounced during a DX pileup or during a contest when you'll hear short exchanges that provide just the minimal amount of information. Some of us like to talk…rag chew…but that can be accomplished with narrowband (SSB) modulation with no problem. It seems that narrowband signals suit our needs. I suppose it would be handy from time to time to be able to send a 3 megabit jpg file to someone I am working on 20m but that's not the main focus of a radio contact.

Of course, not all amateur radio operation is below 1 GHz. There's always someone messing around at microwave and millimeter-wave frequencies. I've done some mountaintop operating at 10 GHz and achieved VUCC on that band. Recently, the ARRL announced a new distance record of 215 km on the 47 GHz band.

ICOM produced a D-STAR system at 1.2 GHz with a data rate of 128 kilobits per second, quite the improvement over AX.25 packet. However, adoption of this technology has been very limited and it remains a single-vendor solution. In fact, it may be a dead technology, hard to say.

There is significant work going on with High-Speed Multimedia (HSMM) Radio which repurposes commercially-available 802.11 ("WiFi") access equipment. Broadband-Hamnet is focused primarily on using 2.4 GHz band to create wireless mesh networks. Amateur Radio Emergency Data Network (AREDN) is doing some interesting work, mostly on the 2.4 GHz and 5.8 GHz bands. The HamWAN site has lots of information about a 5.8 GHz network in the Puget Sound area. I just became aware of the Colorado Amateur Radio Broadband Network, in my neighborhood. The basic theme here is use commercial gear on adjacent ham bands…a common strategy for many VHF and higher ham radio systems.

There are probably some other high-speed digital systems out there that I've missed but these are representative.

Infrastructure Rules

A critical factor in making LTE (and 5G) work is the huge investment in infrastructure by Verizon, AT&T and others. With cellular networks, the range of the radio transmission is limited to a few miles. One of the trends in the industry is toward smaller cells, so that more users can be supported at the highest bandwidths. With 5G moving up in frequency, small cells will become that much more important.

On the other hand, most amateur radio activity is "my radio talking to your radio" without any infrastructure in between. Most of us like the purity and simplicity of my station putting out electromagnetic waves to talk directly to fellow hams. In many cases, this simplicity and robustness has played well under emergency and disaster conditions.

FM (and digital voice) repeaters are a notable exception with the Big Box on the Hill retransmitting our radio signal. For decades now, FM repeaters have represented an infrastructure that individual hams and (more often) radio clubs put in place for use by the local ham community. There is a trend towards more infrastructure dependency in ham radio as repeaters are linked via the internet using IRLP, EchoLink and other systems. (Some hams completely reject any kind of radio activity that relies on established infrastructure, often claiming that it is irrational, unethical or just plain wrong.)

One interesting area that is growing in popularity is the use of hotspots (low power access points) for the digital voice modes (D-STAR, DMR, Fusion, etc.) In this use model, the ham connects a hotspot to their internet connection and talks to anyone on the relevant ham network while walking around the house with a handheld transceiver. See the Brandmeister website to see the extent of this activity. It strikes me that this is the same "small cell" trend that the mobile wireless providers are following. You want good handheld coverage? Stick a hotspot in your house.

Looking at ham radio and broadband communications, I summarize it like this:

- The vast majority of ham radio activity is narrowband, for reasons described above.
- There is some interesting ham radio work being done with broadband systems, mostly on 2.4 GHz and 5.8 GHz.
- Commercially available broadband technology (LTE, 5G, and beyond) will continue to increase total network bandwidth and performance increasing the difference between commercial broadband and narrowband ham radio.

Implications

The reason for writing this article is that the amateur radio community needs to recognize and understand this increasing bandwidth gap. We like to talk about the cool and exciting stuff we do with wireless communications but we need to also appreciate how this is perceived by someone with an LTE phone in their pocket. Just communicating with someone at a distance is no longer novel. After all, Amateur Radio is Not for Talking; see Chapter 20.

What should we do with this? Here's a few options:

1. Don't worry. We are all about narrowband and that's good enough. This attitude might be sufficient as there are tons of fun stuff to do in this narrowband world. In terms of ham radio's future, this implies that we need to expose newcomers to narrowband radio fun. We'll need to get better at talking about how amateur radio makes sense in this broadband world.

2. Embrace commercially available broadband. Use it where it makes sense. This approach means that Part 97 remains mostly narrowband but we can make use of the ever-improving wired and wireless network infrastructure that is available to us.

3. Develop Part 97 ham radio broadband. I am initially a bit skeptical of this idea. How the heck does ham radio compete with the billions of dollars Verizon, AT&T and others poor into broadband wireless? But that may not be the right question. Once again, I fall back to the universal purpose of amateur radio: *To Have Fun Messing Around with Radios*. Can we have fun building out a broadband network? Heck yeah, that sounds like an interesting challenge. Would it be useful? Maybe. Emergency communications might be an appropriate focus and some hams are already working on that. Create a network that operates independent from the commercial internet and make it as resilient as possible. It doesn't have to be at 5G speeds but it better be way faster than AX.25.

I think Option #3 is definitely worth considering.

References

ExtremeTech explains: What is LTE? – https://www.extremetech.com/mobile/110711-what-is-lte

Everything You Need to Know About 5G – https://spectrum.ieee.org/video/telecom/wireless/everything-you-need-to-know-about-5g

Broadband Hamnet – http://www.broadband-hamnet.org/

Amateur Radio Emergency Data Network – https://www.aredn.org/

HamWAN – https://hamwan.org/

Colorado Amateur Radio Broadband Network – https://carbbn.org/

BrandMeister Network – https://brandmeister.network/

37. We've Got Some Explaining to Do

There was a fun interaction on twitter the other day about how we represent amateur radio to the general public. It started with this tweet from @FaradayRF:

This refers to an article in the Las Vegas Review-Journal newspaper where the author decided to use the theme of "ham radio is retro" to tell the story of a ham radio gathering at NAB. I really hate it when ham radio gets positioned as "old technology" in the world of awesome wireless stuff. Clearly, some of our technology is dated, but the amateur service includes lots of new technology and experimentation. (Actually, the tone of the article was very positive, so we shouldn't complain too loudly.)

So I replied, along with a few other folks:

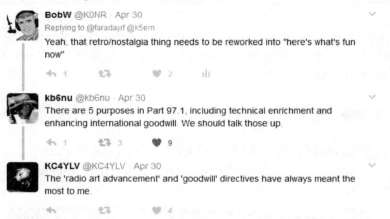

So KB6NU and KC4YLV took the discussion back to good old Part 97 of the FCC rules. (You ever notice how often radio hams like to quote Part 97? It's right up there with the U.S. Constitution and the Declaration of Independence.) I tried to recall from memory the five things listed in 97.1 as the Basis and Purpose of the Amateur Radio Service, but failed.

I had to look them up, so I'll save you the trouble and list them here. Actually, I am going to provide the KØNR Abbreviated Version:

> **Part 97.1 Basis and Purpose of Amateur Radio**
> a) **Voluntary public service, including emergency communications**
> b) **Advancement of the radio art**
> c) **Advancement of communication and technical skills**
> d) **Expansion of trained radio/electronics enthusiasts**
> e) **Enhancement of international good will**

These five things are still relevant and are being pursued today. Not all radio amateurs contribute to every one of these but as a group we are doing these things. The good news is that many non-hams do understand the When All Else Fails aspect of ham radio...most have had their cellphone become a useless brick during major incidents. Items b, c and d are all about learning new things, building skills and expanding the number of radio hams. We should talk more about that. Enhancing international good will may seem a bit quaint but this crazy world can always use another dose of that.

Part 97 does leave out one thing that is the ultimate attraction and, in fact, the universal purpose of ham radio:

> **To Have Fun Messing Around With Radios**

References

FCC Part 97 Rules – http://www.arrl.org/part-97-text

38. VHF FM: The Utility Mode

I've been referring to VHF FM as the *utility mode* for quite a while now. I picked this up from Gary Pearce, KN4AQ, when I inherited the FM column from *CQ VHF* magazine (no longer being published). Gary recently filled me in on the origin of this term, which he captured in his first FM column for *CQ VHF*.

Gary describes how he got hooked on VHF operating, especially 2-meter FM:

> I've been a ham since 1965 (age 15). Today, I have an Extra class license, and I operate some HF (mostly SSB and digital, with cw limited to occasional bouts at Field Day). But since my first days as a Novice with a Heathkit Twoer, I've been a VHFer. I went through the 2 and 6 meter AM days with a Heath Seneca and Utica 650, and then SSB with a Gonset Sidewinder and Hallicrafter HA-2 transverter. But what really flipped my switch as an early ham was an old, single-channel Motorola 80D on 146.94 simplex, installed in the car of a teenage friend's father. For you newer hams, this is an exercise in nostalgia that I don't have space here to explain – I wish I could. I will note that the Motorola 80D was an FM radio that began life in a police car or taxi cab somewhere.

> It was a huge, heavy, all-tube radio that sat in the trunk and improved traction on the ice. Below the dash was a control head with volume, squelch, and the microphone and speaker.
>
> It wasn't long before I learned about repeaters, which enhanced the FM experience immeasurably (all four of them in the Chicago area at the time). My interest in VHF SSB waned...Getting involved in a local repeater group felt comfortable – this was someplace where I could really participate.

But then things shifted as time passed. Gary wrote:

> While I wasn't looking, FM became just another mode. At least that's the consensus I got from some of the guys who have been doing Amateur Radio publishing a lot longer than I have...
>
> Some columns devoted to sub-sets of Amateur Radio have lasted for decades. VHF-UHF is one. That's the weak-signal side of VHF, not the FM side. Digital modes go through enough reincarnations to keep interest up. DX, contesting, QRP, holding their own.
>
> But not FM/Repeaters? QST editor Steve Ford, WB8IMY, suggested why, and gave me the idea for this column's "Utility Mode" tag line. He said, "Our research has shown that while FM users comprise a very large portion of the amateur community, the majority tend to perceive their FM activity more as a 'utility' function rather than a hobby."

VHF FM is arguably the most common mode used in amateur radio. (Can I back that up with reliable data? Not sure, but I'm pretty sure it's true.) I do see where it fits into the concept of a *utility function* or *utility mode*. Think about the electrical system in your house (a utility). For the most part, you just plug things in and use it but you probably don't consider yourself a 120 VAC hobbyist. Well, a few of you might but that's another issue. VHF FM is a lot like that...most hams have it and they just use it without too much consideration. Push the button and it works.

But that definition is a little bit derogatory...VHF FM is just there and no one appreciates it. The Eeyore of ham radio modes.

Another definition of *utility* (as an adjective) is:

> **Utility:** having or made for a number of useful or practical purposes rather than a single, specialized one:
> *a utility knife.*

This fits my perception of VHF FM: very useful for many things. Whether you are providing communications for a bike race, handling talk-in for a local hamfest, working the ISS, chatting across town while mobile, the first choice is likely to be 2-meter FM.

And that's why I've always been a VHF FM enthusiast: there are so many things you can do with it. Just use your imagination.

39. Pursue Radio Operating Goals

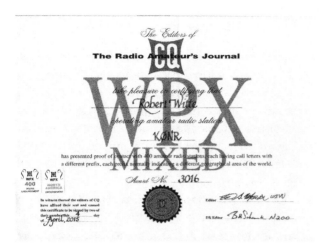

Operating goals or awards are a fun way to keep focused on accomplishing something via ham radio. Really, it's a specific reason to get on the air and make radio contacts. I am not big on idle chit chat via the radio ("the weather here is 65 deg and raining") so having a reason to make contacts helps me get on the air. I've tended to pursue awards in a serial manner...once I hit some level of accomplishment, I usually declare victory and move on to something else.

Way back in the wayback machine, the first award I pursued was Worked All States (WAS). It does take some effort but I was pretty active on the HF bands at the time, so many of the states just showed up in my log. But to really drive it home, I kept track of which states I still needed and actively looked for opportunities to work them.

Next up was Worked All Continents (WAC), which obviously requires working some DX. But then I decided that if I was going to have any DX cred at all, I needed to get DX Century Club (DXCC). This turned out to be a bit of a challenge with my modest station (100 watts and a dipole) but I found that working DX contests to be very helpful. The big hassle was collecting the QSL cards and getting them checked by the ARRL (back before the Logbook of the World was a thing). Once I checked the box on DXCC at just over 100 countries, I was satisfied and went on to other things. Serious DXers chase all available countries/entities to get Honor Roll and other bragging rights.

The VHF and higher bands have always been a passion for me, so I pursued the VHF/UHF Century Club (VUCC) awards. First, it was 6-meter VUCC, the easiest one to get. A really good run during the ARRL June VHF contest can produce the 100 grids you need for the award in one weekend. For me, it took a few more contests than that after factoring in the fallout that occurs when trying to get confirmation QSLs. The 10 GHz VUCC only requires 5 grids which turned out to be not too difficult. My VHF collaborator at the time, Doug/WØAH (now K4LY) and I took turns operating from Pikes Peak while the other guy went out and activated the required 5 grids. It helps to have a big honkin' mountain nearby to use for 10 GHz operating. About this time, I got into working the LEO satellites and worked the required 100 grids for satellite VUCC. I still don't have very many grids confirmed on 2 meters, so that one is still calling to me.

Recently, I spent some effort going for the CQ WPX Award (worked prefixes award). This is an intriguing award structure because every new call sign *prefix* counts as a new one. For the basic mixed-mode award, you need to work at least 400 different call sign prefixes. I found this format to be a lot of fun because "everyone is DX" so to speak, but DX prefixes are also very valuable. This scoring approach is used for the WPX contests, which naturally brings out stations with less common call sign prefixes. A big motivator for me was when the ARRL announced Logbook of the World (LoTW) support for the CQ WPX awards. I mean, there was no way I was going to collect 400 QSL cards to submit for this award, but using LoTW made this very efficient.

Summits On The Air

Lately, I've been active in the Summits On The Air (SOTA) program, both activating and chasing summits. This is a natural fit for me as I've enjoyed mountaintop operating in various forms, mostly on VHF and UHF. The SOTA program has a wide variety of awards, supported by a very powerful database used to record SOTA radio contacts and keep track of the scores. It is not really a competition but there is a friendly rivalry between SOTA enthusiasts as they monitor each other's posted scores.

Position	Activator Callsign	Summits	Points	Seasonal Bonus	Total Score	Avg. points per Expedition	View Log
1	KX0R	1120	4537	543	5080	4.54	View
2	WG0AT	392	1653	375	2028	5.17	View
3	K5RHD	232	1799	192	1991	8.58	View
4	K7PX	318	1537	345	1882	5.92	View
5	N0TA	357	1592	201	1793	5.02	View
6	W0CCA	218	1663	75	1738	7.97	View
7	WA6MM	246	1553	144	1697	6.90	View
8	W0CP	246	1436	174	1610	6.54	View
9	K0MOS	252	1370	186	1556	6.17	View
10	K0JQZ	280	1298	213	1511	5.40	View
11	KD0YOB	311	1297	54	1351	4.34	View
12	KC0YQF	191	1010	156	1166	6.10	View
13	KC5CW	202	1105	36	1141	5.65	View
14	K0NR	172	956	84	1040	6.05	View
15	K0JJW	131	624	57	681	5.20	View
16	AD0KE	82	500	81	581	7.09	View
17	W0MNA	92	497	0	497	5.40	View
18	W0ERI	89	491	0	491	5.52	View
19	AE0AX	77	405	54	459	5.96	View
20	N0BCB	78	356	36	392	5.03	View
21	N6UHB	48	268	57	325	6.77	View
22	N0MTN	61	278	33	311	5.10	View
23	KD0PNK	43	219	60	279	6.49	View
24	K0ZV	41	223	33	256	6.24	View
25	W0ADV	23	218	15	233	10.13	View
26	K0BH	32	210	0	210	6.56	View
27	W4XEN	32	136	27	163	5.09	View
28	K0FTC	37	152	3	155	4.19	View
29	KI0G	27	93	15	108	4.00	View
30	W0ASB	19	91	9	100	5.26	View

Colorado SOTA Activator Scores – Sept 2019

I've been using VHF (and UHF) exclusively for SOTA and managed to qualify for the Shack Sloth Award using just those bands. (Shack Sloth is achieved with 1000 chaser points.) Shack Sloth is a bit of a misnomer for me as many of my SOTA chasing contacts were done

Pursue Radio Operating Goals | 165

while hiking, mobile or portable (not sitting at home in a shack). I recently completed 1000 activator points using only VHF to qualify for the Mountain Goat Award.

The table shows the scores for the Colorado (W0C) SOTA activators (as of Sept 2019). At the top of the list, we find Carey/KXØR totally killing it with over 5000 points. These folks have all reached the coveted Mountain Goat status: KXØR, WGØAT, K5RHD, K7PX, NØTA, WØCCA, WA6MM, WØCP, KØMOS, KØJQZ, KDØYOB, KCØYQF and KC5CW.

Other Goals

You may not find the awards and goals I've mentioned to be very interesting, but there are many other options. In 2016, the ARRL sponsored the National Parks On The Air (NPOTA) program, which created a lot of interest and activity. I did just three activations for NPOTA but many people really got into it.

You might also set your own personal goal, not associated with any award. I know one ham that decided his goal was to make a ham radio contact every day of the entire year. This sounds simple but if you have a full-time job and other responsibilities, it takes some persistence to make this happen. Perhaps you are public service oriented; you might set a goal for the number of ARES events you support this year. I challenge you to think about what it is you are trying to do with ham radio and set a goal that is consistent with that.

40. The VHF Digital Cacophony Continues

Graphic courtesy of HamRadioSchool.com

Wouldn't it be cool if we had one digital communications format for the VHF/UHF amateur bands with all equipment manufacturers offering compatible products? The basic modulation and transport protocol would be standard with manufacturers and experimenters able to innovate on top of that basic capability. There would be plenty of room to compete based on special features but all radios would interoperate at a basic level. You know, kind of like analog FM.

Yeah, we don't have that.

We have three main competing (incompatible) standards in the running for Digital Voice (DV): D-STAR, DMR and Yaesu System Fusion (YSF). At a high level, these three formats all do the same thing but there are significant differences in implementation. All three of these are (arguably) open standards, allowing anyone to implement equipment that supports the standard. However, the reality is that D-STAR is still largely an ICOM system (with Kenwood now joining the party). YSF is mostly a Yaesu system and DMR is...well, DMR is not deeply embraced by any large amateur radio equipment supplier. Instead,

DMR is promoted heavily by Motorola for the commercial market via their MOTOTRBO product line. Another big factor is the availability of DMR radios from some of the low cost providers in the ham market: Anytone, Connect Systems and Tytera.

D-STAR has a clear head start versus the other DV standards and is well-entrenched across the US and around the world. DMR and YSF are the latecomers that are quickly catching up. To put some numbers on the adoption of DV technology, I took a look at the digital repeater listings in the August issue of the Southeastern Repeater Association (SERA) Repeater Journal. SERA is the coordinating body for Georgia, Kentucky, Mississippi, North Carolina, South Carolina, Tennessee, Virginia and West Virginia. This is a large region that includes rural and large urban areas, so perhaps it is a good proxy for the rest of the country. I just considered the listings for D-STAR, DMR and YSF repeaters, some of which are set up as mixed-mode analog and digital repeaters.

```
D-STAR    161    39%
DMR       136    33%
YSF       121    29%
Total:    418    100%
SERA Repeater Journal - August 2016
```

I was definitely surprised at how the DMR and YSF numbers are in the ballpark with D-STAR. Of course, we don't know for sure how many of these repeaters are actually on the air or how many users are active on each one. Still, pretty impressive numbers. (And I did not bother to count the analog FM repeaters but those numbers are way higher, of course.)

It is the repeater clubs and repeater owners that drive the deployment of infrastructure for new technology. To some extent, they are driven by what their users want but also by their own technical interests and biases. One of the positive factors for DMR is that many of these systems are Motorola MOTOTRBO. Hams involved in commercial land mobile radio are exposed to that technology and naturally port it into the amateur radio world. MOTOTRBO is actually not that expensive and it's built for commercial use. More recently, many DMR repeaters have been put on the air using homebrew adaptions of existing radio gear. YSF received a big boost when Yaesu offered their repeater for $500 to clubs and owners that would put them on the air. By using Yaesu's mixed analog/digital mode, it was an easy and attractive upgrade for aging repeater equipment.

Disruption From New Players

Early on in the world of D-STAR, the DV Dongle and DV Access Point by Robin AA4RC allowed hams to access the D-STAR network without needing a local repeater. This basic idea has continued and evolved in several different directions. For example, the DV4Mini is a cute little USB stick that implements a hot spot for...wait for it...D-STAR, DMR and YSF. This is very affordable technology (darn right cheap) that lets any ham develop his or her own local infrastructure. *We don't need no stinkin' repeater.* DV MEGA is another hot spot, supporting D-STAR, DMR and YSF. Oh, and then there's openSPOT...don't want to leave them out. I guess somebody forgot to tell these guys they have to choose one format and religiously support only that one.

All of the popular amateur digital voice (DV) systems (D-STAR, DMR and YSF) use the AMBE vocoder (voice codec) technology. This technology was developed by Digital Voice Systems, Inc. and is proprietary technology covered by various patents. The use of proprietary technology on the ham bands causes some folks to get worked up about it, especially proponents of an open source world. Codec2 is an alternative open voice codec developed by David Rowe, VK5DGR. David is doing some excellent work in this space, which has already produced an open codec that is being used on the ham bands. FreeDV is an umbrella term for this open codec work.

It will be interesting to see if and how Codec2 gets adopted in a DV world already dominated by AMBE. After all, a new codec is another contributor to the digital cacophony. On the HF bands, it is easier to adopt a new mode if it can be implemented via a soundcard interface (which FreeDV can do). Any two hams can load up the right software and start having a QSO. The same is true for weak-signal VHF/UHF via simplex. VHF/UHF repeaters are trickier because you must have a solution for both the infrastructure (repeaters and networks) as well as the user radios.

The majority of digital repeaters support just one digital format. For example, a D-STAR repeater does not usually repeat DMR or YSF transmissions. Interestingly, DMR and YSF repeaters often support analog FM via mixed mode operation for backward compatibility. It is definitely possible to support multiple digital formats in one repeater, but the question is will large numbers of repeater owners/operators choose to do that? With existing DV systems, the networking of repeaters is unique to each format which represents another barrier to interchangeability. Recently, some hams have been experimenting with DV systems that cross-connected different DV technologies. That is, the system supports D-STAR, DMR and YSF radios talking to each other. Very clever.

In the case of a new vocoder, we can think of that as just a new format of bits being transported by the existing DV protocol. DMR, for example, does not actually specify a particular vocoder, it's just that the manufacturers developing DMR equipment have chosen to use AMBE technology. So from a technical viewpoint, it is easy to imagine dropping a new vocoder into the user radio and having it work with other identical radios. Of course, these radios would be incompatible with the existing installed base. Or would they? Perhaps we'd have a backwards compatibility mode that supports communication with the older radios. This is another example of putting more flexibility into the user radio to compensate for DV incompatibilities.

A potential advantage of Codec2 is superior performance at very low signal-to-noise ratio. We've all experienced the not-too-graceful breakup of existing DV transmissions when signals get weak. Some of the Codec2 implementations have shown significant improvement over AMBE at low signal levels.

Conclusions

- For the foreseeable future, we will have D-STAR, DMR and YSF technologies being used in amateur radio. I don't see one of them dominating or any of them disappearing any time soon.
- Equipment that handles all three of those DV modes will be highly desirable. It is the most obvious way to deal with the multiple formats. Software-defined radios will play a key role here.
- A wild card here is DMR. It benefits from being a commercial land mobile standard, so high quality infrastructure equipment is available (both new and used gear). And DMR is being embraced by both land mobile providers (i.e., Motorola, Hytera) and suppliers of low cost radios (i.e., Anytone, Tytera, Connect Systems). This combination may prove to be very powerful.
- Codec2 will struggle to displace the proprietary AMBE vocoder, which is well-established and works. The open source folks will promote codec2 but it will take more than that to get it into widespread use. Perhaps superior performance at low signal levels will make the difference.
- Repeater owner/operators will continue to deploy single-DV-format repeaters. However, keep an eye out for systems that handle multiple formats, especially network servers that tie different repeater and hotspot technologies together.

Well, those are my thoughts on the topic. I wish the DV world was less fragmented but I don't see that changing any time soon.

References

D-STAR (Wikipedia) – https://en.wikipedia.org/wiki/D-STAR

DMR (Wikipedia) – https://en.wikipedia.org/wiki/Digital_mobile_radio

Yaesu System Fusion (YSF) – http://www.yaesu.com/pdf/System_Fusion_text.pdf

PART III
SUMMITS ON THE AIR (SOTA)

Summits On The Air (SOTA) is an award program for radio amateurs that encourages portable operation in mountainous areas. Activating and chasing SOTA summits is an excellent combination of ham radio, hiking/climbing and outdoor exploration.

41. How To Do A VHF SOTA Activation

The *Summits On The Air* (SOTA) program has really taken off in North America. SOTA originated in the UK in 2002 and it took a little while for it to make it across the Atlantic to this continent. The basic idea of SOTA is to operate from a designated list of summits or to work other radio operators when they activate the summits. The list of designated summits are assigned scoring points based on elevation and there are scoring systems for both *activators* (radio operators on a summit) and *chasers* (radio operators working someone on a summit).

Handheld transceiver with half-wave antenna for 2 meters.

Most of the operating is on the HF bands but there are quite a few VHF contacts on SOTA. Obviously, HF has the advantage of being able to work longer distances without too much trouble. Typically, the HF station is your classic portable QRP rig, portable antenna and battery power. (A portable power source is required and the use of fossil fuels is prohibited.) Being a VHF enthusiast, I prefer the challenge of making contacts above 50 MHz, so my SOTA contacts are usually on the 2-meter and 70-centimeter bands.

My basic VHF SOTA station is a handheld FM transceiver with a half-wave telescoping antenna. The standard rubber duck on a handheld transceiver (HT) is generally a poor radiator so using a half-wave antenna is a huge improvement. This simple station is an easy addition to my normal hiking routine…just stuff the HT and antenna in my backpack along with the usual hiking essentials and head for the summit.

To receive activation points for SOTA, you need to make a minimum of 4 contacts from the summit. If I am hiking a summit within range of a major city, I can usually just make some random contacts by calling CQ on the National Simplex Calling Frequency, 146.52 MHz. However, operating in more remote areas requires a little more planning. I'd hate to hike all that way and come up short on the required contacts, so I use a few tactics to rustle up some VHF contacts. Of course, I will post my planned activation on the SOTAwatch site in advance, to let people know that I'll be on the air. While this goes out worldwide, it may not reach the right radio amateurs within VHF range. The next thing I do is send an email to some of VHF-equipped hams I know will be within range. Many people respond to such a request to work a summit, even if they are not active in SOTA. When on the summit, my first call is on 146.52 MHz or some other popular simplex frequency. If I don't raise anyone there, I will make a call on a few of the 2m repeaters in the area to see if someone will come over to "five two" to make a contact. SOTA does not recognize repeater contacts but it is OK to solicit simplex contacts using a repeater. These techniques and a little patience have always gotten me at least four contacts, and usually quite a few more.

Bob/KØNR operating with 2m Yagi antenna on Mt Sneffels.

The omnidirectional antenna of the basic VHF SOTA station will make some contacts, adding some antenna gain can really help your signal. There are a number of compact directional antennas that are easy to take hiking. Elk Antennas makes a log-periodic antenna that covers 2 meters and 70 cm. Another popular antenna is the 2 meter / 70 cm Yagi antenna made by Arrow Antenna. These antennas are lightweight and assemble/dissemble easily, which is important to hiking radio operators.

So far, most of the SOTA VHF activity in North America is on on 2-meter FM, the *utility mode*. Everyone seems to have a 2-meter HT, so tossing it in a backpack and heading out is a natural thing to do. Using my FT-817, I have made some VHF contacts on CW and SSB. These modes are much more efficient than FM and the station on the other end is usually well-equipped. Nothing like a big gun station with huge antennas to help pull your QRP signal out of the noise! I expect the use of CW and SSB to increase on VHF as SOTA becomes more popular. While FM activity uses vertical polarization (antenna elements are vertical), most SSB/CW activity uses horizontal polarization (antenna elements are horizontal).

Summits On The Air is a great way to take ham radio outdoors. So get off the couch, find a summit and have some fun with ham radio.

References

Summits On The Air (Official Site) – https://sota.org.uk/

SOTA Introduction to Activating (John/G4YSS) – https://sota.org.uk/Joining-In/Introduction-to-Activating

42. How To Do a SOTA Activation On Pikes Peak

Perhaps this should be called *The Slacker's Guide to Activating Pikes Peak* since I am going to describe the easy way to do a Summits On The Air (SOTA) activation on America's Mountain. If you plan to hike up, you have my complete support but this post is not meant for you.

The summit of Pikes Peak. (Photo: Ken Wyatt/ WA6TTY)

Pikes Peak (W0C/FR-004) is about 10 miles straight west of downtown Colorado Springs. At an elevation of 14,115 feet, the mountain towers over Colorado Springs and the other front range cities. (You may see the elevation listed as 14,110 but it was revised upward in 2002 by the USGS.) This means that it has an excellent radio horizon to large populated areas. On VHF, it is common to work stations in Kansas, Nebraska, Wyoming and New Mexico. (See Chapter 17: VHF Distance From Pikes Peak.) On HF, you'll do even better.

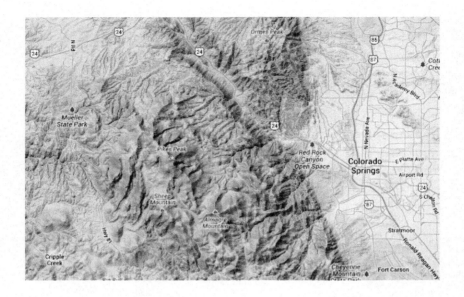

Access to the summit has three options: hike up, drive up via the Pikes Peak Highway or ride the Pikes Peak Cog Railway. Most people will probably choose the highway since the cog railway only gives you 30 to 40 minutes on the summit. (Normally, you return on the same train that takes you to the top. You can try to schedule two one-way trips but that is a challenge.)

Note: in 2019, the Pikes Peak Cog Railway is out of service but is expected to re-open in 2021.

The highway is at a well-marked exit off Highway 24, west of Colorado Springs. There is a "toll" to use the highway (~$15 per person, check the Pikes Peak website for details.) The road is now paved all of the way to the top and is usually in good shape. The only caution on driving up is that *some people* get freaked out by sections of the road that have steep drop-offs without guard rails. It is very safe but I know some folks just can't handle it. The main caution driving down is to use low gear and stay off your brakes. There are plenty of signs reminding you to do this and during the summer there is a brake check station at Glen Cove where the rangers check the temperature of your brakes.

A new summit house is being constructed at the summit of Pikes Peak, expected to be completed in 2020. Access to the summit is somewhat limited and a free shuttle service takes you the last few miles. Check the Pikes Peak website for the current status.

It takes about an hour to drive to the summit, assuming you don't dawdle. It is best to drive up during the morning and avoid the afternoon thunderstorms. (Check the Pikes Peak website for when the gate opens, usually 7:30 am in the summer. Or call 719 385-7325 for a recorded message on road conditions.) Once you get to the summit, you'll find a large circular parking area, the summit house and a few other buildings. The summit of Pikes is broad, flat and rocky, so pick out a spot away from the buildings for your SOTA adventure. There are quite a few radio transmitters on the peak so expect some interference. Since this is way above treeline, your antennas will have to be self supporting. For VHF, giving a call on 146.52 MHz FM will usually get you a few contacts and sometimes a bit of a pileup. Be aware that on top of Pikes you are *hearing everyone* but they can't always hear each other. It can get confusing. Some other VHF simplex frequencies worth trying are 147.42 MHz (The Colorado 14er Event frequency) and 146.46 MHz (a local 2-meter hangout frequency). If you have 2-meter SSB, call on 144.200 MHz USB. On the HF bands, pray for good ionospheric conditions and do your normal SOTA thing.

Your body and your brain will likely be moving a little slower at 14,000 feet due to the lack of oxygen. Don't be surprised if you have trouble deciphering and logging call signs. Take it slow and monitor your physical condition on the peak.

Bring warm, layered clothes even in the summer, since Pikes Peak can have arctic conditions any time of the year. Keep a close eye on the weather since thunderstorms are quite common during the summer months. Lightning is a very real danger, so abandon the peak before the storms arrive.

References

Pikes Peak website
http://pikespeak.us.com/

Pikes Peak (W0C/FR-004)
https://summits.sota.org.uk/summit/W0C/FR-004

Pikes Peak Highway
https://coloradosprings.gov/pikes-peak-americas-mountain

43. Mt Herman: SOTA plus VHF Contest

In September 2014, the North America SOTA Weekend coincided with the ARRL September VHF Contest, which I interpreted as a great opportunity to do a combination SOTA activation and QRP VHF operation. A few other folks thought that was a good idea so we all got on the air from SOTA peaks on the Sunday of the weekend. I decided to operate from Mount Herman (W0C/FR-063) in grid DM79. I hiked up the same mountain for the previous year's September contest and got soaked by the rain. Fortunately, the weather was excellent this year, making it a great day.

The view from the summit of Mount Herman.

For radio equipment, I took a couple of HTs for 2m and 70cm FM and the FT-817 for CW/SSB on 6m, 2m and 70cm. Most of the SOTA action would be on 2m FM but SSB is critical for working the VHF contest. I did put out the word to the usual VHF contesters that there would be FM activity and did work a few of them via 2m FM. The 2m FM calling frequency, 146.52 MHz, is commonly used for SOTA but is not allowed for contest use. (Another example of how this rule is just a barrier to contest activity.) We used 146.55 MHz for the contest contacts. [Note: the ARRL now allows the use of 146.52 MHz in contests.]

The Yaesu FT-817 transceiver in operation.

I had coordinated with Brad/WA6MM who was going to be on Grays Peak (W0C/FR-002), one of the Colorado 14ers. When he made the summit, I had my 2-meter Yagi antenna pointed in his direction and easily worked him on 2-meter FM at a distance of 65 miles. Brad was using an HT with a half-wave vertical antenna. Also, I worked Stu/W0STU and Dan/N0OLD on Bald Mountain (W0C/FR-093), which sits on the east side of I-25 right at Monument Hill. Contest activity was light, as usual for the September contest in Colorado. We did have two rover stations that activated a few of the unpopulated grids in eastern Colorado: George/AB0YM and Jonesy/W3DHJ.

```
Band   QSOs  X pt  =QSO pts.  X  Grids  =  Points
-------------------------------------------------
 50      8    1        8          5          40
144     23    1       23          5         115
432     14    2       28          3          84
-------------------------------------------------
TOTALS  45            59         13         767
```

My contest score was not bad for a few hours of operating QRP portable. It turns out that I had set the Colorado section record for "single-op portable" back in 1990 with just 624 points (using my old call sign KBØCY). Oddly enough, 24 years later it appears that I set a new record. (This speaks more to the lack of QRP activity during the September contest and less about my incredible operating ability.)

All in all, it was a great day in the mountains to take a hike and play with radios. I will probably do the SOTA + VHF Contest activation again.

44. The Ten Essentials for Hiking (and SOTA Activations)

Most backcountry hikers are familiar with the Ten Essentials that you should take with you whenever you head into the wilderness. Over the past few years, I noticed that I was getting a bit sloppy with regard to what is actually in my pack when I head out on the trail. This hit home one day when my GPS battery went dead. I fumbled around to find my compass which *was supposed to be* in my pack. Well, it was in my pack, the other one that I left at home.

This caused me to review the list of ten essentials to make sure I had the right stuff in my kit. A search on the internet revealed that the classic list of ten has been modified and augmented by various people to make it better. (Innovation runs rampant on the interwebz, you know.) One of the better resources I found was a page on the REI website, which explains how the Classic Ten Essentials have been updated to the Ten Essential *Systems:*

1. **Navigation (map and compass)**
2. **Sun protection (sunglasses and sunscreen)**
3. **Insulation (extra clothing)**
4. **Illumination (headlamp/flashlight)**
5. **First-aid supplies**
6. **Fire (waterproof matches/lighter/candles)**
7. **Repair kit and tools**
8. **Nutrition (extra food)**
9. **Hydration (extra water)**
10. **Emergency shelter**

Read through the REI web page to get the fine points of this system approach. I won't repeat that information here. They also include a *Beyond the Top Ten* list which calls out the need for:

Communication device: Two-way radios, a cell phone or a satellite telephone can add a measure of safety in many situations.

Of course, what they really mean is an amateur radio transceiver and antenna but they probably can't say that in print due to licensing issues. (Not everyone in the backcountry has an FCC ham license. I know, they all *should* have an amateur license but many don't...very hard to understand.)

So how are you doing with your *Ten Essentials* list? Are you consistent in taking along the right stuff in your pack?

References

REI Ten Essentials
http://www.rei.com/learn/expert-advice/ten-essentials.html

45. The Most Radio-Active Mountain in Colorado

KØNR Operating VHF on Mt Herman (Photo: Ken Wyatt, WA6TTY)

I've often said that Mount Herman (W0C/FR-063) is the most (ham) radio-active mountain in Colorado. Many of us have operated from that summit for VHF contests, QRP events and Summits On The Air (SOTA). A review of the SOTA database for activations in Colorado through July 14, 2019 shows these four summits are the most-activated SOTA peaks:

Summit	SOTA Designator	Number of Activations
Mount Herman	W0C/FR-063	217
Pikes Peak	W0C/FR-004	111
Genesee Mountain	W0C/FR-194	102
Mount Evans	W0C/FR-003	75

All of these summits are relatively close to the large population centers in the state. Also, they are not that difficult to get to and some of them have roads that go to the top. Pikes and Evans are both 14ers but can be accessed via 2WD vehicles.

But what makes Mount Herman so special? It does *not* have a road to the summit — you definitely have to hike it, a little bit more than a mile one way with elevation gain of ~1000 feet. What makes the difference for *The Hermanator* is that it is in the backyard of the well-known radio ham, goat hiker and SOTA enthusiast, Steve/WGØAT. Steve has personally activated the summit more than 78 times AND he ~~drags along~~ encourages other radio hams to join him. Fortunately for me, Mt Herman is about 4 miles as the GPS flies from my house, so I have worked that summit 65 times (usually Steve and always on VHF). I've also activated Mt Herman for SOTA 7 times.

Steve/WGØAT, thanks for all of the Qs from Mt Herman over the past years! See you on the air and on the trail.

References

Mount Herman SOTA (W0C/FR-063)
https://www.sota.org.uk/Summit/W0C/FR-063

46. Yaesu FT-1DR: A Trail Friendly SOTA HT

A common topic in the QRP community is the *Trail Friendly Radio (TFR)* concept for backpack-style operating on the high-frequency bands. I've adapted the concept for the VHF/UHF bands, calling it the *VHF Trail Friendly Radio (VTFR)*.

The Yaesu FT-1DR is a trail-friendly radio for VHF.

Strong candidates for the best VTFR include the Elecraft KX3 (with 2m option) and the Yaesu FT-817. Heck, both of these radios deliver all of the HF bands, 6m and 2m in one portable package. (The FT-817 also has 70cm.)

But the other set of strong contenders for the best VTFR is one of the many dual-band HTs available on the market. It is hard to beat the compact, portable attributes of these great little radios for casual use on the trail. I'm not going to review them all but instead talk about my current favorite: the Yaesu FT-1DR. (Yaesu has recently replaced the FT-1DR with the newer model FT1XDR, which is the same design but with an improved GPS receiver and larger battery pack.)

My main usage of the radio is when hiking and doing Summits On The Air (SOTA) activations. This radio has a lot to offer in terms of capability and features, but the main things that stand out are 2m/70cm band coverage, two independent receivers and built-in GPS/APRS capability. Most SOTA VHF operating is on 2m FM so that band is critical, but I also make contacts on 70cm. More important is that together 2m and 70cm covers that vast majority of FM repeaters in my state, providing the best backcountry repeater coverage. The built-in APRS features allow the HT to be an effective tracking device as I move down the trail. SOTA chasers can see my position in real-time and anticipate when I'll be on the summit. The radio has two separate receivers which turns out to be very useful when on the trail. With two receivers, I can monitor 146.52 MHz while also keeping an ear on a local 2m or 70cm repeater. Another configuration is using one side of the radio to ping my location via APRS while the other side monitors 146.52 MHz.

The extended receive capability of the radio opens up lots of listening options: AM broadcast, FM broadcast, airband, shortwave and NOAA weather radio. I don't use these very often but there are times that I want to tune to weather or news.

I am not a huge fan of Yaesu's C4FM digital mode but do use it on occasion. The DN (digital narrow) mode supports voice and position information simultaneously, so Joyce/KØJJW and have been using it to keep track of each other on the trail. The radio provides a basic indication of distance and direction to another C4FM radio.

A few other tips: if you buy an FT-1DR, I recommend upgrading the belt clip to the BC-102 clip from Batteries America. It is way better than the standard one from Yaesu. Michael/KX6A created a very handy quick reference card for operating the FT-1D, so consider putting one in your pack.

Note: The Yaesu FT-1DR is no longer sold but you can find it on the used market. Also, check out more recent radios with similar features such as the Yaesu FT-3DR.

References

BC-102 Belt Clip
http://www.batteriesamerica.com/yaesu-vertex.htm

KX6A Quick Reference Card (FT-1DR)
http://mwgblog.com/archives/2016/01/05/yaesu-ft1d-sota-trail-card/

47. SOTA plus NPOTA on Signal Mountain

Signal Mountain (W7Y/TT-161) is now my favorite radio location in the Grand Teton National Park. The summit is well-marked on the Grand Teton NP map, on the east side of Jackson Lake. It has a paved road to the top and it provides excellent views of Jackson Hole and the surrounding mountains. Oh, and it's a great location for ham radio.

On this summit, I did a combination Summits On The Air (SOTA) and National Parks On the Air (NPOTA) activation. Well, sort of. It turns out that when I packed for the trip, I included my usual SOTA gear, which is all VHF. For NPOTA, I loaded up my HF DXpedition gear that needs a pretty hefty power source. These means that the HF stuff uses my car battery, so it is not SOTA-compliant. Oh well.

For the SOTA activation, I used the Yaesu FT-1DR and my 3-element Arrow yagi antenna to work a handful of stations on 146.52 MHz. I was a little concerned about finding enough stations listening on 52, but once again a little bit of patience paid off and I made my four QSOs.

Then I set up the NPOTA station to activate Grand Teton National Park (NP23). My equipment was a Yaesu FT-991 driving an end-fed half-wave for 20 meters from LNR Precision. I've tried a number of different portable antennas over the years but have found that a half-wave radiator up in the air is a pretty effective antenna. This could be a center-fed dipole antenna but that can be a challenge to support, depending on the physical location.

Bob/K0NR works stations on 2m FM for a SOTA activation.

The end-fed half-waves (EFHW) from LNR Precision are easily supported using a non-conductive pole such as the 10m SOTABEAMS pole. The top two sections of the pole are too thin to support much of an antenna, so I have removed them. This makes my pole about 9 meters in length which is still long enough to support a 20-meter half wave. (The antenna angles out a bit as shown in the photo but it's pretty much vertical.) I attached the pole to a fence post using some hook/loop straps. I don't fiddle with the length of the antenna, I just let the antenna tuner in the FT-991 trim up the match. This is the same configuration I used in Antigua (V29RW), where it worked great.

Bob/K0NR using the "back of the SUV" operating position.

The FT-991 is a great little radio for this kind of operation. The SUV we were driving is not set up for HF operation so I just located the radio in the back of the vehicle and plopped down on a folding camp chair. For power, I clipped directly onto the vehicle battery with fused 10 gauge wires.

I started by making a few calls on 20-meter SSB. As soon as I was "spotted" on the usual websites, I had a good pileup going. I worked 40 stations in about 40 minutes, so averaged one QSO per minute overall. Thanks to everyone that worked me; all contacts have been uploaded to Logbook of The World.

Oh, and it was a lot of fun.

References

Signal Mountain SOTA (W7Y/TT-161)
http://www.sota.org.uk/Summit/W7Y/TT-161

48. Monarch Ridge South SOTA Activation

For the 2016 Colorado 14er Event, I chose really easy SOTA summits to activate because I had fractured my ankle earlier in the summer. At this point, I was able to hobble around with a protective boot but walking more than a few hundred feet was difficult. On Saturday, our radio crew (Bob/KØNR, Joyce/KØJJW, Denny/KB9DPF and Kathy/KB9GVC) drove up Pikes Peak and took a short stroll away from the vehicle to operate. On Sunday, we decided to activate Monarch Ridge South (W0C/SP-058), using the sight-seeing tram that goes to the top.

The trailhead sign at Monarch Pass.

Access is right off Monarch Pass (Hwy 50), where there is a large parking lot. There is a trail that goes to the top and we'll be back to hike that some other time. The trail is a popular mountain bike route, part of the Continental Divide Trail (CTD). Monarch Pass is oriented north/

south and the trail heads off to the east (behind the tram building). At some point, you need to turn left off the main trail to ascend the summit.

But we opted for the tram, boarding it inside this building near the parking lot. We purchased tickets in the nearby gift shop, which is worth a look if you need a map, book, ice cream cone or trinkets. See their website for latest schedule and pricing.

The tram headed up the mountain.

Here's a photo of the tram going up the side of the mountain. Of course, the views are great and the ride takes about 10 minutes. The tram car holds four people and a reasonable amount of SOTA gear.

At the top, Denny/KB9DPF made contacts on 2m FM, aided by expert logger Kathy/KB9GVC. The actual summit is about a tenth of a mile to the south but we operated from a concrete pad on the north side. The ridge is flat and we judged the activation area to be very large. We made a total of 13 contacts on 2m and 70cm, including 5 other SOTA summits (S2S).

Denny/ KB9DPF operating on 2 meters while Kathy/ KB9GVC logs the contacts.

I used my Yaesu FT-817 to call on 432.100 MHz SSB, hoping to find someone in the UHF contest that was happening concurrently with the 14er event. I didn't work anyone on 70cm SSB but I did work K3ILC in Colorado Springs on FM at a distance of ~90 miles. Not too bad. The Arrow antenna is attached to my hiking stick via the camera mount thread.

There is a substantial radio site on Monarch Ridge that did provide some RF interference to us on 2 meters. The 70 cm band seemed to be unaffected but I can't be sure. The "bad boy" transmitter is the KMYP automated weather station (AWOS) transmitting continuously on 124.175 MHz. Well, at least we could receive current weather information. We did relocate to put some distance between us and the transmitters but my lack of mobility kept us from going very far.

Bob/K0NR operating 70 cm SSB with a small Yagi antenna.

If you are looking for an easy access SOTA summit near Monarch Pass with excellent views, this is it. The hike up should not be very difficult but the tram makes it even easier. If you plan to operate 2 meters, expect some interference. Next time, I'll try locating even further away from the transmitter site. I might even bring along some bandpass filters. Other SOTA enthusiasts have reported no problems on the HF bands.

49. Smoky Mountain SOTA

Joyce/KØJJW and I were getting prepared for a trip to Gatlinburg, TN in August with some of her family. Gatlinburg is the gateway town to the Great Smoky Mountains National Park and the surrounding area. I had hiked and camped in the Smokies years ago and this was a great opportunity to visit that area again. Of course, we needed to get in a little Summits On The Air (SOTA) action during this trip.

We decided to pick out some easy-to-access summits in the area so we could weave them into the trip without too much disruption. My first step was to consult the SOTA database for potential summits in Tennessee and North Carolina, looking at the summits with the most activations. This is usually a good indication of easy access and not too difficult of a climb. I did pick out two iconic summits to activate: Clingmans Dome (this highest summit in the national park) and Mount Mitchell (the highest summit east of the Mississippi river in the US). After checking the various trip reports logged on the SOTA website, I created a list of potential summits. Clingmans Dome and Mount Mitchell were *Must Do* but any other summits would be more opportunistic based on available time and location.

We use VHF/UHF for SOTA activations and opted for a basic FM station for this trip: a pair of Yaesu FT-1DR handhelds, a couple of vertical antennas and a 3-element Arrow Yagi antenna for 2 meters. I debated about whether to bring along the Yagi but the split-boom design fits into my luggage without any problem. In the end, I am glad we had the extra gain of the yagi as several of the contacts would have been missed without it.

Greentop

Greentop (W4T/SU-076) was our first summit...basically a drive-up mountain with radio towers and a lookout tower on top. I noticed quite a bit of interference on the 2-meter band, something I've encountered in previous activations near transmitter sites.

Use of two handheld radios on Greentop to mitigate strong RF interference.

It turns out that putting a more effective antenna on an HT (such as a half-wave vertical) couples more of the interference into the receiver and degrades its performance. On the other hand, the standard rubber duck antenna picks up less of the interference and performs better then the "good" antenna. After I realized this was happening, I tried using two HTs with reasonable results: one radio with a rubber duck was used for receive on 146.52 MHz while another radio with a half-wave antenna was used for transmitting. The net result was acceptable performance that allowed us to make contacts on 2m FM.

Clingmans Dome

Clingmans Dome (W4C/WM-001) is a popular tourist spot in the Great Smoky Mountain National Park. Parking is a challenge and there are quite a few people on the short trail to the summit.

The observation tower at Clingmans Dome, with lots of tourists.

Although it sits right on the border of Tennessee and North Carolina, it is in the W4C (Carolinas) Association for SOTA purposes. As I approached the summit, I saw a fishing pole sticking up in the air. I thought "huh, I wonder what the rangers are demonstrating today."

Followed by "Hey, wait a minute, that looks like a SOTA activation." Sure enough, I met W2SE and WI2W setting up on 20-meter CW. Joyce and I headed to the observation tower and worked 2 meters from up there. There were quite a few people on the observation tower so I considered just operating from down below. I decided to leave the Yagi in the backpack and just use the half-wave vertical. We fit right in with the chaos of tourists on the tower.

W2SE and WI2W on Clingmans Dome.

Mount Mitchell

At 6684 feet in elevation, Mount Mitchell (W4C/CM-001) is the highest point in the USA east of the Mississippi River. (Interesting perspective: our house in Colorado is 800 feet higher than this summit.) We started with just the 2m vertical but switched to using the yagi when we had trouble copying a few stations. It definitely made a difference...probably 6 dB or so. When signals are near the FM threshold, this can pull them out of the noise.

Summit sign on Mount Mitchell.

One of the highlights on Mitchell was working Kevin/K4KPK on Walnut Mountain (W4T/SU-033), summit to summit. Kevin is very active in SOTA and has contributed many SOTA summit guides in the area. I made good use of these reports when planning our trip. He is also the top activator in the W4G (Georgia) association, a Mountain Goat approaching 2000 points.

Richland Balsam and Waterrock Knob

We discovered a number of summits right along the Blue Ridge Parkway and we ended up working these two: Richland Balsam (W4C/WM-003) and Waterrock Knob (W4C/WM-004). Another flashback for us was driving sections of the parkway, which is a lovely drive (typically 45 MPH speed limit) that winds through the mountains. It has been years since we've been on that road. This route is something I'd like to explore further on a future trip as you could spend a week wandering along the parkway and knocking out summits.

Joyce/K0JJW at the summit sign for Richland Balsam Mountain.

We worked Pat/KI4SVM on 2m FM from Watterock. I recognized his call sign from the trip reports he has submitted to the SOTA website. Later, I looked up his SOTA score and found that he is a double Mountain Goat (> 2000 activation points) and the highest scoring activator in the W4C association.

Brasstown Bald

The Mountain Explorer Award is a SOTA award for activating in different SOTA Associations (regions). Activating in Tennessee (W4T association) and North Carolina (W4C association) got my total to 6. Joyce pointed out that we might be able to also hit Georgia on the trip, so we added Brasstown Bald (W4G/NG-001) to the list. This is the highest summit in Georgia, so it rounded out our collection of state high spots for TN, NC and GA.

Bob/KØNR holding the 2m Yagi antenna while Joyce/KØJJW works the radio.

Brasstown Bald is an easy hike-up summit with a significant observation tower on top, including a visitors center. This is another location where we experienced interference from radio gear on the summit, so we chose our position carefully and used the 2-meter Yagi to point away from the interference sources.

Bob/KØNR operating 70 cm FM from Brasstown Bald.

This trip worked out really well. We managed to activate 6 summits for a total of 58 points, operate from three new SOTA associations (W4T, W4C and W4G), enjoy some really nice hikes and see some great scenery. I was a little concerned whether we would find enough random activity on 2m FM for our SOTA activations but it all worked out. Actually, there were a few times that 146.52 MHz was busy and we had to standby to make a call. Some of our contacts were less than 25 miles but many covered 100 miles or more. Yes, the 3-element Yagi made a difference.

If you are in the Gatlinburg area, it certainly makes sense to try a few SOTA activations. I am also thinking about a return trip to enjoy the area more fully including some longer hikes. We really liked hiking the trails and summits there. The elevation is lower than Colorado (read: you have oxygen to breathe), the forests have lots of deciduous trees (not just evergreens) and the trails are less rocky. I am sure we will be back.

50. Pikes Peak SOTA Winter Activation

Joyce/KØJJW and I had intended to hike Pikes Peak this year for a Summits On The Air (SOTA) activation but somehow the plan never came together. I still had my eye on it as a drive-up activation before the end of 2017. The road to the summit is open year round now but closes frequently due to snowstorms passing through. Saturday morning the road was open to 13 miles (out of 19 miles) with the promise that it would be open to the summit later in the morning. (Call 719 385-7325 for a recorded message on road conditions.) By the time we got to the toll gate around 10 am, the road was open to the summit.

KØJJW and KØNR caught on the Pikes Peak webcam (Photo: Paul Signorelli, WØRW)

When we reached the summit, the weather conditions were 20 deg F with 20 mph winds, creating a windchill of 4 deg F. We were prepared for that having loaded up on the winter clothing. Still, it was freaking cold up there. As you can see in the webcam picture, there was only traces of snow on the summit.

To be SOTA-compliant, we had all of our gear loaded into our packs and walked some distance away from our vehicle to set up. Because of the wind, we chose the observation platform, tucked in behind one of the walls. Normally, that platform is to be avoided because its overrun with tourists but with the cold weather we only had a few people to contend with.

Bob/KØNR hunkered down out of the wind. (Photo: Joyce Witte, KØJJW)

Joyce set up on 2m FM (146.52 MHz) using a handheld transceiver (HT) with a vertical antenna. Even with her headset (foam protection on the microphone), the wind noise on her signal was significant. I started out on 2m FM but quickly moved up to 223.5 MHz and worked a few stations there, then on to 446.0 MHz. I had HTs and small Yagi's on both of those bands. Then I fired up 1.2 GHz with an Alinco HT (just 1 W on that band) and a 16-element Yagi, worked Paul/WØRW, Gary/WB5PJB and Wayne/NØPOH. My QSO with NØPOH in Aurora was a new personal best for distance on 23cm/1.2 GHz, at about 90 km.

I tried 2m SSB using my FT-817 but made only one contact: Jim/WBØGMR. Shortly thereafter, I switched back to 2m FM using the 25 W mini-mobile rig with a 3-element Yagi to work many more stations. Again, just running a bit of power and having a decent antenna on 2m FM was very effective at making radio contacts. I expected the Tytera radio to be overloaded with signals on the summit of Pikes but it actually held up well with just occasional bursts of interference.

Overall, we made 54 QSOs (not too shabby): 43 QSOs on 2m, 5 QSOs on 70cm and 3 QSOs on both 1.25m and 23cm. Our best DX was Jeff/NØXLF near Akron, CO for a distance of about 130 miles (on 2m and 70cm).

51. More Power For VHF SOTA

For years now, I've been doing Summits On The Air (SOTA) activations using VHF and higher frequencies. The "go to" band/mode for VHF SOTA is 2-meter FM because of its overall popularity. Just about everyone has a 2-meter FM radio (well, almost everyone). Still, if you are on a remote peak you may not find anyone within range to work. Because of this, it really helps to optimize the performance of your portable VHF station.

Antennas

I've already written that the first step is to upgrade the rubber duck antenna to something that actually radiates. My measurements indicate that a half-wave antenna performs 8 to 10 dB better than your typical rubber duck. That's a big difference. I tend to favor the collapsible half-wave antennas because they are compact and don't require any support. Another option is the J-pole or Slim Jim antennas, typically build out of twin lead or ladder line.

The next step up is to use a small Yagi antenna, such as the 3-element Arrow antenna. Although Arrow does not specific the gain of this antenna, it has been measured at the Central States VHF Society conference as having ~6 dBd of gain. I've been on the lookout for a higher gain antenna but I have not found one that has significantly more gain while still being backpack portable.

Modulation

Frequency Modulation performs very badly when signals are weak. The well-known threshold effect means as the signal level decreases at the receiver it simply crashes into the noise. Linear modes such as CW and SSB work much better when signals are weak, which is why they are popular with the serious VHF crowd. I've used my Yaesu FT-817 to make SOTA contacts on both 2m and 70cm SSB and CW. My all time best distance on 2m during a SOTA activation was 229 miles, a QSO with N7KA from Capulin Mountain using CW. However, the problem with SSB/CW is that there are fewer radio amateurs that operate that mode. I estimate that on a typical day, there are 10 to 100 times more hams on 2m FM than are on 2m SSB/CW.

More Power on FM

I've noticed that I sometimes hear stations on 2-meter FM but they cannot hear me. Further investigation revealed that they were typically running more power than me. I had my little HT putting out 5 W and they were running a 50 W mobile. That got me thinking about whether I could increase my power while still having a backpack-compatible station. SOTA operation is typically QRP, around 5 or 10 W of power. However, SOTA does not specifically state a required power level...it's really driven by the need to operate backpack portable. Hence, there are very few 1 KW amplifiers in use on SOTA summits.

Tytera TH-8600 2m / 70cm transceiver.

Some of the Chinese manufacturers now offer compact dual-band (and even quad-band) VHF/UHF transceivers that output 10 to 30 W of RF power. I purchased the Tytera TH-8600 based on my experience with other Tytera products. The radio's specified output power is 25 W on 2 meters. The DC power current is rated as 0.2 A on receive and 4 A on transmit, not too bad for battery operation. I paired it with a 13.2V LiFe battery rated at 4300 mAH. In theory, that would provide over an hour of transmit time or 21 hours of receive. That should be plenty for the typical SOTA activation. The size is slightly larger than 4" W x 1.5" H x 5" D and it weighs about 2 pounds. All in all, this setup is very compatible with the typical backpack portable operation.

Let's do a little math to understand the difference in transmit signal. The TH-8600 puts out 25W compared to the 5W from FT-60. The difference in dB is 10 log (25/5) = 7 dB. Someone said to me "Hey, that's only a little more than one S unit, which is normally defined as 6 dB. Is that really enough to make a difference?" To which I responded, "yes, 7 dB can make the difference between making the radio contact or not...*when signals are near the noise floor of the receiver.*" For strong signals, it just doesn't matter.

I've used this configuration on three SOTA activations and I like the results. On two of the activations, I compared the TH-8600 (25W) to the Yaesu FT-60 (5W) that my hiking partner (Joyce/K0JJW) was using. Both radios were connected to 1/2-wave vertical antennas, operating on 2m FM. The radios performed the same on receive, as expected. But the weaker stations we were working had trouble hearing the FT-60. Again, if signals were strong, it didn't matter but the extra power made the difference when near the noise floor.

> **Increase your transmit power for additional punch on VHF.**

I checked out the basic performance of the radio on my test bench and found it to be adequate. The transmit frequency was spot on, the harmonics and spurious on 2m were about 60 dB below the carrier. The receiver sensitivity was about 0.2 microvolts. The RF output power was low, 22.4 W on 2m and 17.7W on 70cm (compared to the specs at 25W and 20W).

I was hoping the receiver performance would be better with regards to rejecting adjacent channel signals and intermodulation. I don't have a good test bench for that but I can tell you that I noticed some unwanted interference from transmitters that were not close to my location.

How Many dBs Is That?

So let's summarize the dB situation by comparing all of the potential improvements to the standard handheld transceiver (HT) with a rubber duck antenna. Note that the Yagi gain is specified as dB relative to a half-wave dipole, which has roughly the same performance as a half-wave vertical.

```
5W HT with standard rubber duck antenna      0 dB
5W HT with half-wave antenna                 +8 dB
3-element Yagi antenna (Arrow or similar)    +6 dB
25W transceiver (vs 5W output)               +7 dB
Total improvement (25W yagi vs HT/duck)      +21 dB
```

Wow, I can improve my signal strength by over 20 dB by making these improvements! I should point out that the antenna improvements help on both transmit and receive, while the increased transmit power only improves your stations transmitted signal.

My conclusion is that this type of mini-transceiver can be a good way to go for 2-meter FM SOTA without adding too much of a load in my pack. I expect that I'll still do some HT-only activations but the higher power option is very useful on remote peaks.

Some folks have found my table of dB calculations to be confusing, so here's some more info. I arbitrarily started with the 5W HT and rubber duck at 0 dB. My measurements in the past showed that a half-wave vertical is 8 to 10 dB better than a rubber duck. I decided to use the 8 dB number…it is not a precise measurement anyway and will vary with the specific duck antenna. So that means that the half-wave vertical is +8 dB relative to the rubber duck. The Yagi gain is about 6 dB relative to a dipole (6 dBd). The standard dipole is a 1/2-wave radiator and performs roughly the same as the 1/2-wave vertical, so we'll consider them equivalent. That is, the yagi is 6 dB better than the 1/2-wave vertical. Finally, the 25 W power vs 5 W power adds in 7 dB. Add them all together and you get 0 + 8 + 6 + 7 = 21 dB.

52. Operating Tips for VHF SOTA

Here are some operating tips that I have found useful when doing Summits On The Air (SOTA) on VHF. Joyce/KØJJW and I have been using VHF and higher frequencies for SOTA exclusively and have activated over 100 summits in Colorado (and other states such as California, Wyoming, New Mexico, North Carolina and Tennessee.)

The challenge with VHF and higher is that the radio range is limited compared with HF. (It really does help to bounce those signals off the ionosphere.) VHF propagation will vary depending on a lot of factors but for SOTA activations our range is typically 50 to 100 miles. In the backcountry of Colorado, a 50 mile radius may not include very many active radio amateurs, so you may come up short in terms of radio contacts. For your first activation(s), you may want to stay close to a metropolitan area.

First, take a look at my blog posting about the basics: Chapter 41: How To Do a VHF SOTA Activation

Next, here are some additional tips to having a successful activation:

Send Invitations

Get the word out to people that may be within range. I try to keep track of who I've worked in the past or know to be in a particular area and let them know when I'll be activating. In some cases, I'll go ahead and make a specific sked with a station. That is, we'll meet on a particular frequency at a particular time, perhaps on CW or SSB for maximum weak signal performance.

Create an *Alert* and then *Spot* yourself on the sotawatch.org web page (using smartphone app).

Make Your Call

By the rules, SOTA contacts are always simplex, so on 2 meters the place to try is usually 146.52 MHz, the *National Simplex Calling Frequency*. (Some areas have established other 2-meter simplex frequencies to use for SOTA, so inquire locally.) This is the place to try calling.

Who are you going to find there? Hard to say. SOTA chasers will often listen to "Five Two"…that's where they find SOTA activators. Also, there are radio hams that just like to hang out on 52…they like 2-meter FM operating but they don't like repeaters so they listen here. You will also find that many mobile stations monitor 146.52 as they drive through rural areas. Basically, they are listening for anyone around but aren't flipping from repeater to repeater as they change location. I'll often hear 4WD enthusiasts, hikers, boaters, campers, etc. using Five Two.

We usually say something short and sweet when operating FM repeaters ("KØNR Monitoring") but for SOTA we need to make a longer call. Realize that not everyone knows about SOTA so they may not know what the heck you are doing. Make a call such as this: "CQ CQ this is Kilo Zero November Romeo operating from Pikes Peak – Summits On The Air – anyone around?" Or "CQ 2 meters this is KØNR on Pikes Peak – Summits On The Air." Do a little bit of "selling" when you make your call. Sound like you are having fun.

Sometimes people have their radios set to scan multiple frequencies with 146.52 MHz set as one of the channels. If your transmission is too short, they will miss it. Also, it is helpful to mention the frequency you are calling on…sometimes I'll say "CQ Five Two This is Kilo Zero November Romeo…"

Be Patient

Be patient. Sometimes it just takes time to accumulate your 4 QSOs to qualify for activator points. The few times I have been skunked on QSOs were when I did not have enough time, usually because storms were moving in. Keep calling every few minutes, taking a break once in a while to enjoy the view.

Try Other Frequencies

You may want to try other simplex frequencies, so know the band plan for the area you are activating in. Another trick is to get on a local 2-meter repeater and ask for someone to move over to simplex to work you. I have not had to do this very often but it does work. Have some of the local repeaters programmed into your radio.

Most of the SOTA VHF activity is on the 2-meter band, so that will be the "Go To" band for most activations. However, it is fun to try other bands such as 6 meters and 70 centimeters. I've also been playing around with 1.2 GHz (23 centimeters).

Improve Your Station

Hopefully by now you know that using an HT with a rubber duck antenna is a bad idea. Get at least a half-wave vertical or roll-up j-pole antenna. Better yet, get a small Yagi antenna to add 5 or 6 dB to your signal. It can make a big difference on who you can work. I've even started carrying a small 2m/70cm transceiver to get my 2-meter transmit power up to 25 W. The objective is to extend your operating range.

Another angle is to use SSB and CW, which are much better for weak-signal performance (compared to FM). This sets you up to work the serious VHF stations which tend to use these modes and have large high-gain antennas (horizontally polarized). Think about extending the range of your communications from a 50 mile radius to 200 miles, what a difference! (My best 2m distance on SOTA is 229 miles, using 2-meter CW on Capulin Mountain.)

53. VHF/UHF Omni Antenna for SOTA Use

For Summits On The Air (SOTA), I've been using just the VHF/UHF ham bands, with the 2-meter band being the most popular. For most activations, I use a 3-element Arrow II Yagi antenna that has a gain of about 6 dBd. Sometimes that extra gain makes the difference between completing a contact or not.

Omnidirectional Antenna

But it is also handy to have an omnidirectional antenna that is easy to deploy. Sometimes I'd rather just call or monitor using an omni without having to point the antenna. "Easy and good enough" can be an effective strategy for SOTA.

I usually carry one of the TWAYRDIO RH770 VHF/UHF antennas for use with my 2m/70cm handheld transceiver. Despite its low cost, I have found that its performance to be good. That antenna is offered with a variety of connectors, including a BNC.

> **A key advantage of an omnidirectional antenna is that it is always pointed in the right direction.**

The RH770 2m/70cm vertical antenna for portable use.

This led me to the idea of putting together a simple antenna mount with a BNC on it, attach an RH770 antenna to it and support it using some kind of pole. I have several monopod devices (intended for use as a camera support) that use the standard 1/4-20 thread. I also have a trekking pole that has the same camera mounting thread. Another option is to use an actual camera tripod which is a bit bulky but may work for some SOTA activations.

After a short visit to the hardware store, I selected a crossbar for mounting a light fixture that was about the right size and shape. I happened to have a bulkhead-mount BNC-to-BNC connector which I inserted into the large hole in the crossbar. That hole was not quite large enough for the connector, but a few minutes of work with a round file solved that problem.

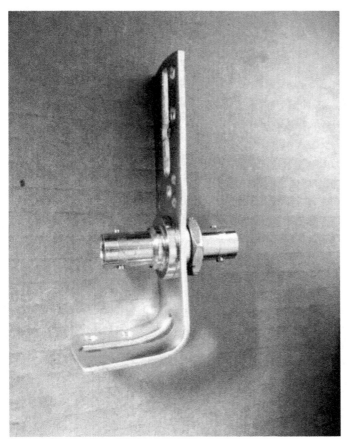

The mounting bracket with BNC bulkhead connector installed.

The crossbar was originally flat but I bent one end of it 90 degrees, with the idea that this might offer other mounting configurations in the future. For example, I might be able to strap or tape the 90-degree angle member to a pole or support.

The monopole with 1/4-20 camera mount thread pokes through the mount.

The other end of the crossbar has a large slot that accommodated the 1/4-20 mounting stud. Actually, the slot was not quite wide enough, so some addition work with a round file opened it up. I secured the 1/4-20 thread using a nylon wing nut.

A wing nut holds the mount onto the monopole.

I've used this setup on several SOTA activations and I am pleased with the results. I carried the crossbar mount attached to the monopod in my pack. On the summit, I simply installed the RH770 antenna onto the top BNC and extended the monopod. On this summit, I found the perfect pile of rocks that made a good support for the monopod. Then I used a short length of RG-8X coax between the bottom BNC and the 2m/70cm transceiver.

The vertical antenna in use, mounted on a monopod stick.

Although my primary interest was with the 2-meter band, it was really convenient to have both 2m and 70cm on the same antenna. I am pleased with the operation of the antenna and the ability to deploy it quickly. I expect to carry this on most of my SOTA activations.

Glossary

ATV Amateur Television, the transmission of video signals via amateur radio.

ARES Amateur Radio Emergency Service, an organization that provides public service and emergency communications via amateur radio.

ARRL American Radio Relay League, also known as the National Association for Amateur Radio

AM Amplitude Modulation, a form of modulation that varies the amplitude of the radio frequency carrier based on the modulating signal.

Autopatch a device that interfaces a repeater to the telephone system, permitting radio amateurs to make telephone calls via the repeater.

Break the term used to interrupt a conversation, normally reserved for priority or emergency traffic.

Courtesy Beep the audible beep (or other signal) that occurs after the repeater's timeout timer is reset. Repeater users should pause between transmissions to let this reset occur and to let others break in.

CTCSS Continuous-Tone-Coded Squelch System, subaudible tones used for accessing some repeaters. These tones are in the frequency range of 67 Hertz to 250.3 Hertz.

CW Continous Wave, on-off modulation of a carrier using Morse Code.

dB decibel, a logarithmic method for expressing the ratio of two numbers. A doubling of signal power corresponds to 3 dB.

Digipeater a digital packet repeater for retransmitting packet radio signals.

DTMF Tones Dual-Tone Multi-Frequency tones which are produced by pressing a telephone or radio keypad (otherwise known as Touch-tones, which is an AT&T trademark).

Duplex operation using a pair of frequencies, one for transmit and one for receive, as when using a repeater.

EME Earth-Moon-Earth, also known as moonbounce, the technique of bouncing signals off the moon to conduct radio communications between two points on earth.

FM Frequency Modulation, a modulation technique that places information on a transmitted signal by modulating (varying) the frequency.

Full Quieting a strong received signal having no noise in it.

HAAT Height Above Average Terrain is a measure of how high a radio antenna is above the surrounding landscape.

HT – handheld transceiver (sometimes "Handie-Talkie" a trademarked Motorola name).

Input Frequency the frequency that a repeater listens on (and the frequency that a repeater user transmits on).

Kerchunk to key a repeater without identifying your station, often followed by the offending transceiver bursting into flames.

LoTW Logbook of The World, the ARRL-sponsored amateur radio logging database.

Machine slang for repeater or repeater system.

Output Frequency the frequency that a repeater transmits on (and the frequency that a repeater user listens on).

OSCAR Orbiting Satellite Carrying Amateur Radio, amateur radio satellites operating in space (usually with transponders or repeaters)

PL Private Line, the Motorola trademarked name for CTCSS.

RACES Radio Amateur Civil Emergency Service, an emergency communications group operating under a special section of the Amateur Service regulations.

SSB Single-sideband modulation, an efficient form of amplitude modulation (AM) that only uses one sideband

Simplex radio communications using the same transmit and receive frequency (as in communication between two stations without the use of a repeater).

Transmit Offset the difference between the repeater user's transmit and receive frequencies. This offset is either + or – 600 kHz on most 2-meter repeaters.

UHF Ultra High Frequency, 300 MHz to 3 GHz.

VHF Very High Frequency, 30 to 300 MHz.

WSJT modes Weak Signal Joe Taylor modes that use advanced digital signal processing for excellent weak-signal performance. These modes include JT65, MSK144, FT8, etc. Invented by Joe Taylor, W1JT.

WSPR Weak Signal Propagation Reporter, a system for sending and receiving low-power transmissions to test propagation paths on the HF bands.

Made in the USA
Middletown, DE
04 December 2019